The Four Laws That Do Not Drive The Universe

Elements of Thermodynamics
for the Curious and Intelligent

The Four Laws That Do Not Drive The Universe

Elements of Thermodynamics
for the Curious and Intelligent

Arieh Ben-Naim

The Hebrew University of Jerusalem, Israel

World Scientific

NEW JERSEY • LONDON • SINGAPORE • BEIJING • SHANGHAI • HONG KONG • TAIPEI • CHENNAI • TOKYO

Published by

World Scientific Publishing Co. Pte. Ltd.
5 Toh Tuck Link, Singapore 596224
USA office: 27 Warren Street, Suite 401-402, Hackensack, NJ 07601
UK office: 57 Shelton Street, Covent Garden, London WC2H 9HE

Library of Congress Cataloging-in-Publication Data
Names: Ben-Naim, Arieh, 1934– author.
Title: The four laws that do not drive the universe : elements of
   thermodynamics for the curious and intelligent / Arieh Ben-Naim,
   The Hebrew University of Jerusalem, Israel.
Other titles: Elements of thermodynamics for the curious and intelligent
Description: New Jersey : World Scientific, 2017. |
   Includes bibliographical references and index.
Identifiers: LCCN 2017023316| ISBN 9789813223479 (hardcover : alk. paper) |
   ISBN 9813223472 (hardcover : alk. paper) |
   ISBN 9789813223486 (pbk. : alk. paper) |
   ISBN 9813223480 (pbk. : alk. paper)
Subjects: LCSH: Thermodynamics.
Classification: LCC QC311 .B3947 2017 | DDC 536/.7--dc23
LC record available at https://lccn.loc.gov/2017023316

British Library Cataloguing-in-Publication Data
A catalogue record for this book is available from the British Library.

Copyright © 2018 by World Scientific Publishing Co. Pte. Ltd.

All rights reserved. This book, or parts thereof, may not be reproduced in any form or by any means, electronic or mechanical, including photocopying, recording or any information storage and retrieval system now known or to be invented, without written permission from the publisher.

For photocopying of material in this volume, please pay a copying fee through the Copyright Clearance Center, Inc., 222 Rosewood Drive, Danvers, MA 01923, USA. In this case permission to photocopy is not required from the publisher.

Printed in Singapore

This book is dedicated to all those who read or who will read Atkins's book:

FOUR LAWS THAT DRIVE THE UNIVERSE

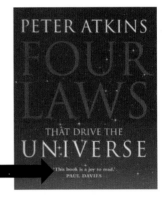

This book is a joy to read
Paul Davies

פֶּתִי, יַאֲמִין לְכָל-דָּבָר; וְעָרוּם, יָבִין לַאֲשֻׁרוֹ.
נָחֲלוּ פְתָאיִם אִוֶּלֶת; וַעֲרוּמִים, יַכְתִּרוּ דָעַת.
משלי, יד; טו,יח

Contents

List of Abbreviations   xi
Preface   xiii
Acknowledgements   xvii

1 Introduction: Basic Concepts of Thermodynamics   1
    1.1 Macroscopic Characterization of a Thermodynamic System   2
    1.2 Thermodynamic Parameters   5
    1.3 The Existence of Equilibrium States   11
    1.4 Thermodynamic States and State Functions   16
    1.5 The Structure of Thermodynamics   18
    1.6 A Brief Introduction to Shannon's Measure of Information (SMI)   25
       1.6.1 The Four Steps of Obtaining Entropy from the SMI   26
       1.6.2 First Step: Playing the Uniform 20Q Game   28
       1.6.3 Second Step: Playing the Non-uniform 20Q Game   31
       1.6.4 The Third Step: Generalization from a 20Q Game to an Over 20^{23}Q Game   34

viii *The Four Laws That Do Not Drive The Universe*

      1.6.5 The Final Step: The Entropy Spawns as a Particular Case of a 20Q Game   37
  1.7 Some Concluding Remarks   38

2 The Zeroth Law (ZL)   41

  2.1 The Thermal Zeroth Law   42
  2.2 The Mechanical Zeroth Law   47
  2.3 The Material Zeroth Law   48
  2.4 The Law of Existence of Equilibrium States of a Thermodynamic System   50

3 The First Law (FL)   53

  3.1 Conservation of Energy in Simple Physical Systems   54
  3.2 The Extension of the Law of Conservation of Energy to Include Thermal Energy   56
  3.3 The End of the Two Separate Laws; the Conservation of Energy, and the Conservation of Matter   61
  3.4 Applications of the FL   63
  3.5 Some Final Thoughts on the First Law   64

4 The Second Law (SL)   69

  4.1 Introduction   72
  4.2 The Various Formulations of the SL   73
  4.3 Probability and Probability Distribution   75
  4.4 The Probability Formulation of the Second Law   85

	4.4.1 Generalizations of the Probability Formulation of the Second Law	88
4.5	The Entropy Formulation of the Second Law	93
	4.5.1 Application of the SMI to a Thermodynamic System	94
	4.5.2 Definition of the SMI on the Probability Distribution of Locations and Velocities of the Particles	95
	4.5.3 Definition of Entropy	96
	4.5.4 The Entropy Formulation of the Second Law	99
4.6	The Helmholtz Energy Formulation of the Second Law	106
4.7	The Gibbs Energy Formulation of the Second Law	115
4.8	Applications of the Second Law	118
	4.8.1 A System Having N Solvent Molecules and One Simple Solute Particle s	119
	4.8.2 A System with N Solvent Molecules and One Solute Particle s Having One Internal Rotational Degree of Freedom	121
4.9	Some Final Thoughts on the Second Law	128

5 The Third Law (TL)   135

5.1	The Various Formulation of the TL	136
5.2	Applications of the Third Law	138
5.3	Adiabatic Cooling	143

	5.4 Concluding Remarks	149
6	Which of the Four Laws Drive the Universe?	151

Notes 159
References and Suggested Reading 175
Index 179

List of Abbreviations

Av	Avogadro
FL	First Law
PD	Probability distribution
Pr	Probability
20Q	Twenty questions
SMI	Shannon's Measure of Information
SL	Second Law
TE	Thermal Equilibrium
TL	Third Law
T4L	Atkins' book "Four Laws that Drive the Universe"
ZL	Zeroth Law
Av	Avogadro

Preface

This book is addressed to any curious and intelligent reader. The book aims to tickle, and hopefully to satisfy your curiosity. It also aims to challenge your gray matter, and to enrich your knowledge by telling you some facts and ideas regarding the Four Laws of Thermodynamics.

I am aware of the possibility that potential readers of this book might have already been exposed to Atkins' book with a similar title. Truth be told, I hope very much that you have read that book, and if you haven't, I encourage you to read it. You will be rewarded by comparing the two books.

This book challenges both the title, as well as the contents of Atkins' (2007) book, *Four Laws that Drive the Universe* (T4L). One can glean from the title of my book my diametrically opposed views from Atkins' posture. Here is how Atkins' book is described on its jacket:

> *Peter Atkins explains the basis and deeper implications of each law, drawing out their precision, clarity, and beauty… the unstoppable rise of entropy explains why*

> *your desk tends to get messier and why the Universe must one day die — an outcome of the iconic Second Law which C.P. Snow famously argued should be as familiar to any educated person as the works of Shakespeare.*

This is a totally misleading message to the potential reader of Atkins' book, T4L. My desk does not tend to get messier, and if it were to do so, it would have nothing to do with the Second Law. The so-called "unstoppable rise of entropy" *does not explain* anything! Also, the author does not have license to predict the death of the universe either one day or one night. I will further comment about Atkins' misleading statements throughout this book.

Although one does not need any mathematics in order to understand the elements of thermodynamics, a certain measure of mathematics is indispensable in order to derive some of the consequences of the four laws, and their applications. The level of mathematics required in thermodynamics is not particularly high, however it is within this apparently low-level of mathematics where pitfalls lie, waiting to swallow their prey. This is the main reason so many authors have shamelessly fallen into this trap.

Thus, while the book is written in a simple, friendly and non-mathematical fashion, the reader is advised to consult one of the technical books listed in the bibliography in order to understand the book's content.

I also believe that researchers and teachers of thermodynamics will benefit even more from reading this book than the lay reader who has never been exposed to thermodynamics. The reason behind this is that those who have had prior exposure to, or written textbooks, or taught thermodynamics might need to invest greater effort in "Un-learning" and disabusing themselves of the subject as it is presented in other textbooks, and "convert" to the new approach, and a new way of looking at the Four Laws of Thermodynamics.

This book is organized in five chapters. The first discusses some basic concepts used in thermodynamics. Those who are familiar with the elements of thermodynamics may skip Sections 1.1 to 1.5 of Chapter 1, and read Section 1.6 only.

The succeeding four chapters discuss the Four Laws. Each chapter starts with a brief description of the law, and then discusses the various formulations and some applications of the law. Finally, a few questions are posed which are not normally raised in textbooks on thermodynamics such as:

Is the law absolute or are there any exceptions? How does the law rank in comparison with the other laws of thermodynamics? Can we do without this law? Can we expect the law to be valid for billions of years from now, in either the past, or in the future? And finally, the intriguing question: Does the law drive anything; a simple experiment, life's processes, or the entire universe?

I hope that you will read the book carefully and critically so that you will be able to answer these questions by yourself before moving on to read my suggested answers. In any case, any comments from you, the readers, will be welcomed.

Arieh Ben-Naim

Department of Physical Chemistry
The Hebrew University of Jerusalem
Jerusalem, Israel
Email: ariehbook@gmail.com
URL: ariehbennaim.com

Acknowledgements

I am grateful to John Anderson, Robert Engel, Jose Angel Sordo Gonzalo, Robert Hanlon, Shannon Hunter, Richard Henchman, Zvi Kirson, Bernard Lavenda, Mike Rainbolt, and Steven Bottomley for reading parts or the entire manuscript and offering useful comments. Thanks also to Alex Vaisman for some of his drawings used in this book.

As always, I am very grateful for the gracious help I got from my wife, Ruby, and for her unwavering involvement in every stage of the writing, typing, editing, re-editing and polishing the book.

Sensing Hot and Cold

1

Introduction: Basic Concepts of Thermodynamics

Thermodynamics deals with the characterization of macroscopic systems and the changes that occur in such systems. This chapter is devoted to describing and defining the various thermodynamic systems and the parameters with which we characterize such systems. In most of this book we shall discuss the laws of thermodynamics for the simplest systems. These systems consist of *one component*, say, pure argon in the gaseous phase, pure water in the liquid state, or pure iron in the solid state. We also assume that there are no external fields that operate on the system, such as electrical, magnetic, or gravitational. Clearly, there is no system which is not affected by any external field. However, we can think of an ideal system for which such effects are negligible. We shall discuss the concept of a *thermodynamic equilibrium*, and the meaning of a *state function*. Finally, we shall present a very brief introduction to Shannon's Measure of Information, a concept which is indispensable for understanding the meaning of entropy and the Second Law.

1.1 MACROSCOPIC CHARACTERIZATION OF A THERMODYNAMIC SYSTEM

A macroscopic thermodynamic system, which we will later refer to simply as a *system*, is a system consisting of a huge number of small particles, atoms, or molecules. In the early 20th century some textbooks referred to the existence of atoms and molecules as a *postulate*. Nowadays, this "postulate" is a well-accepted *fact*, and is usually not used as one of the postulates in characterizing a thermodynamic system. However, we should add that when we talk about a macroscopic system, we assume that it consists of a huge number of particles, typically of the order $N_{Av} \cong 6.023 \times 10^{23}$ particles. N_{Av} is referred to as Avogadro's number. We shall see in Chapter 4 why it is *necessary* to assume that the number of particles is of that order of magnitude.

We shall distinguish between several types of systems:

(a) *Isolated system.* This is a system which does not exchange anything with its surroundings.

It is usually assumed that it has a fixed volume (V), energy (E), and number of particles (N). We shall also assume that such a system is not affected by any external field. Of course, such an idealized isolated system does not exist. However, it is convenient to assume the existence of such a system for building up both thermodynamics and statistical thermodynamics. Thus, we can view an isolated system as one having solid (inflexible) walls (fixing its volume), which are perfect insulators (not allowing flow of heat — see Chapter 3), and impermeable (not allowing exchange of matter). We shall refer to such an isolated system as an (E, V, N)-system, meaning a system having a fixed energy, volume and number of particles.

Note that we use the notation (E, V, N) for either the choice of the independent thermodynamic variables, or for characterization of a thermodynamic system, say an isolated system for which these variables have fixed values.

(b) *Adiabatic system.* A system having perfectly insulating walls that do not allow the exchange of heat between the system and its surroundings.

(c) *Open and closed system.* We usually refer to a *closed* system as one which does not exchange matter with its surroundings. An *open* system can exchange matter with its surroundings. Sometimes, when the system contains different kinds of atoms or molecules, we can further classify systems as being *closed* to one or more kinds of molecules, but *open* to flow of other kinds of molecules.

(d) *Isothermal system.* This is a system held by a thermostat at constant temperature (T). In most cases we also assume that the system has a fixed volume (V), and fixed number of particles (N). In such a case we shall refer to such a system as a (T, V, N)-system.

(e) *Isothermal-isobaric system.* This is a system having a constant temperature (T), and a constant pressure (P). Usually, such a system is held by a thermostat and sealed with a movable piston that maintains a constant pressure. Such a system is referred to as a (T, P, N)-system, (Fig. 1.1).

I hope that you now have an idea about the type of systems we will be dealing with in this book. Let us proceed to discuss the parameters with which we characterize thermodynamic systems.

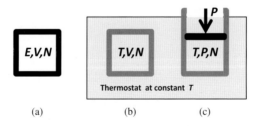

Figure 1.1 Three commonly discussed thermodynamic systems: (a) An isolated system, characterized by (E, V, N). (b) An isothermal system characterized by (T, V, N). (c) An isothermal-isobaric system characterized by (T, P, N).

1.2 THERMODYNAMIC PARAMETERS

We assume that the reader is familiar with concepts such as; temperature, pressure, density, volume and energy, etc. We shall later define new concepts such as entropy, Gibbs energy, etc. It is convenient to distinguish between two classes of parameters, or thermodynamic variables.

(i) *Extensive Parameters*

These parameters depend on the size of the system. More specifically, if we double the system the corresponding parameter is doubled. Examples are the volume, the total number of particles, etc. We shall encounter other parameters later on. Figure 1.2 shows two systems having the same volume V. The combined system has volume $2V$. Similarly, if the number of molecules in each system is N, then the number of molecules in the combined system is $2N$.

Thus, the extensive parameters are *additive*. It should be noted that this additive property is valid

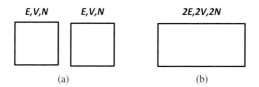

Figure 1.2 Additivity of an extensive quantity. (a) Two identical systems, each having the same E, V and N. (b) The combined system have doubled volume ($2\,V$), doubled energy ($2\,E$) and doubled number of particles ($2\,N$).

provided that no process occurs when we combine the two systems. Taking 100 cm³ of sugar and one liter of water, the total volume of the combined systems would be 1100 cm³. However, if we allow the two systems to mix the sugar will dissolve in the water, and the total volume will, in general, not be the sum of the two volumes. In fact, if the solubility of the sugar is larger than 100 cm³/liter of water, then the volume of the water will be nearly unchanged upon the addition of sugar, (Fig. 1.3a). In other cases, if we take 100 cm³ of salt which is very soluble in water, the total volume of the combined

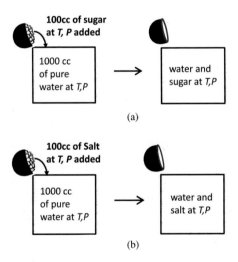

Figure 1.3 (a) Small volume of sugar (volume $v = 100$cc) added to pure water of volume $V = 1000$cc, at T, P. The new system's volume might be different from 1100cc. Cc = Cubic centimeter. (b) Small volume of salt (volume $v = 100$cc) added to pure water of volume $V = 1000$cc, at T, P. The new system's volume might be different from 1100cc.

system, after the salt dissolved might be even *smaller* than the initial volume of water, (Fig. 1.3b). Another interesting case is the addition of water to a dilute solution of ethanol in water. In this case adding 100 cm^3 of water will cause an increase of the volume of more than 100 cm^3.

(ii) *Intensive Parameters*

An intensive parameter is a "non-additive" parameter. These are parameters that do not sum when we combine two systems. For instance, if we combine two systems having the same T, P, and N, the combined system will have the same T, the same P but doubled N, (Fig. 1.4). Again, one should be careful to specify under which conditions we *combine* the systems. Examples of intensive parameters are temperature, pressure, density, etc.

Normally, for a single phase system at equilibrium the intensive parameters have a constant value at each point in the system, and they do not change with time. For instance, if we measure the temperature of the system at equilibrium at different points we should get the same temperature. The same is true for multi-phase systems.

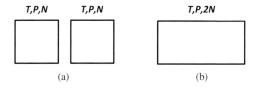

Figure 1.4 The characteristics of an intensive variable: (a) Two identical systems having the same T, P and N. (b) The combined system has the same T and P but doubled N.

However, in a two phase system, say, liquid water and solid ice, the *density* of water in the two phases at equilibrium will be quite different. In any case, if we double the system, the intensive parameters will not be doubled. So far we have discussed qualitatively the parameters with which we describe a thermodynamic system. We have used the term "equilibrium" without defining it. In fact, we do not have a clear cut definition of the concept of equilibrium. We shall discuss this in the next section.

The three most important intensive parameters are the temperature, the pressure, and the chemical potential. As we shall see in the next chapter these parameters "control" the "flow" of an extensive quantity in the following sense: When there is a difference in temperatures between two bodies, and when heat can flow between them (diathermal wall), then heat will flow from the high to the low temperature, (Fig. 1.5).

When there is a pressure difference between two systems connected by a movable barrier (or a piston), then volume will "flow" from the system with the low, to the high pressure. We do not usually speak of a "flow of volume." What we mean by "flow of volume" is that the volume of the system having the higher pressure will

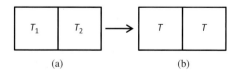

Figure 1.5 If $T1 > T2$, then heat will flow from the high temperature body to the low temperature body. At equilibrium the temperatures of the two bodies will be equal.

Introduction: Basic Concepts of Thermodynamics 9

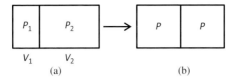

Figure 1.6 If $P_1 > P_2$, then volume will "flow" from the low pressure compartment to the high pressure compartment. *i.e.*, V_1 will increase and V_2 will decrease.

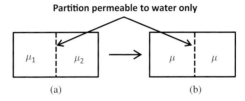

Figure 1.7 Two compartments containing water and sugar. μ is the chemical potential of water. If $\mu_1 > \mu_2$, then water molecules will flow from the high to the low chemical potential compartment. The flow will stop when the chemical potentials of water in the two compartments become equal.

increase, and the volume of the system having the lower pressure will be reduced, see (Fig. 1.6).

The relatively new and important concept defined in thermodynamics is the chemical potential. We shall not define it here. We shall say only that, as the temperature "controls" the flow of heat, and the pressure "controls" the flow of volume, the chemical potential "controls" the flow of matter. In fact, there are different chemical potentials associated with different compounds. Also, the "flow" of a specific compound can be realized in two senses. Figure 1.7 shows two compartments containing water and sugar, separated by a partition which is permeable to water but not to the sugar. In this case, water will flow from the system having higher water chemical potential to the lower water chemical potential. Similarly, the partition could be permeable to the sugar but not to the water. In this case, the flow of sugar will be from the high sugar chemical potential to the low sugar chemical potential. These two flows can be said to be spatial flow (sometimes referred to as diffusive flow). There is another "flow" of matter which is not spatial. A simple example is shown in Fig. 1.8. Initially, we have a pure 1,2 dichloroethene, (Fig. 1.8a), in a system at equilibrium. We add a catalyst (which is the analogue of the permeable partition in Fig. 1.7), and we observe that some molecules will "flow" from the cis-isomer to the trans-isomer, (Fig. 1.8b). This flow will stop when the chemical potentials of the two isomers are equal.

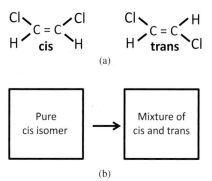

Figure 1.8 (a) Two isomers of 1,2 dichloroethene; cis and trans. (b) On the left side we have pure 1,2 dichloroethene. When a catalyst is added there will be a "flow" from the cis-isomer to the trans-isomer. At equilibrium there will be a mixture of the cis- and trans-isomers having the same chemical potential.

1.3 THE EXISTENCE OF EQUILIBRIUM STATES

A central concept in thermodynamics is the *state of equilibrium*. As used in thermodynamics this concept is similar to, but not exactly the same as in physics. In physics, the term equilibrium is used in two senses: one, when there are two opposing forces acting on something so that the net effect is "no effect." The simplest case is shown in Fig. 1.9a. If the two forces acting on the hard sphere are equal in magnitude but opposite in directions, then the two forces cancel each other, and the sphere will stay at its location.

The second case is when a ball is at the bottom of a potential well, (Fig. 1.9b). In this case the equilibrium state means that the state — the location of the ball, is *stable*. If it is slightly moved to the right or left in Fig. 1.9b (or in any other direction in three-dimensional case), a force is generated to restore the ball to its original location. Such a state is called a stable state.

In both cases the word "stable" features in the description of the equilibrium state. This word will also feature in the thermodynamic equilibrium described below.

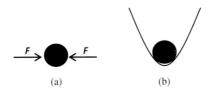

Figure 1.9 (a) Two equal and opposing forces, F, acting on a ball. (b) A ball at a minimum of a potential function.

Consider a system of N hard particles confined to a box of volume V, and having total energy E. For simplicity, we shall always assume that the energy of the particles is the *kinetic* energy of translation. (Note: a particle having mass m, and velocity v, has kinetic energy of $mv^2/2$). The particles are moving incessantly, occasionally colliding with each other and with the wall.

A *classical description* of the *state* of each particle consists of its location R, and its velocity v. Similarly, the microscopic classical state of N particles is completely specified by the locations R_i, and the velocities v_i of all the N particles. In general, real atoms and molecules have *internal degrees* of *freedom* and the corresponding internal states, such as rotational, vibrational, electronic, and nuclear energies. In the present discussion we neglect all these. Not because they are small, or *negligible*, but because they do not change in the processes we will be discussing in most of this book. Of course, in some chemical processes, or a nuclear reaction we must take into account these internal states of the particles.

It is an experimental fact that after some time, most thermodynamic systems will reach a state we call an *equilibrium* state, which can be characterized by a small number of thermodynamics parameters. For instance, a system consisting of one mole of argon (i.e. N_{Av} atoms of argons) can be described by its temperature T and volume V.

Obviously, the characterization of a system by three variables, say, T, V, N, is far from a microscopic description of the same system which requires $3N$ locational and $3N$ velocity coordinates. The fact that the same system can be described by a handful of parameters is far from being trivial. This thermodynamic description means that if you know the three quantities T, N, V all other thermodynamic parameters will be *determined* by these parameters. For instance, the well-known equation of state of an ideal gas provides us with the pressure of the gas described by T, N, V. Explicitly, given T, N, V we can calculate the pressure by: $P = k_B NT/V$, where k_B is the Boltzmann constant $k_B = 1.38 \times 10^{-23} J K^{-1}$. Similarly, the energy, the entropy, the Gibbs energy, etc. are determined once we know the three parameters T, N, V.

In addition, thermodynamics provides a host of thermodynamic relationships between the thermodynamic quantities of the same system, such as an equation for the heat capacity in terms of the temperature derivative of the entropy and many others.

All these relationships are valid only when the system is at an equilibrium state. If we take N atoms in an ideal gas confined to a volume V, and having a temperature T, we can tell its pressure from the equation $P = k_B NT/V$. Now, suppose that we remove the partition between the two compartments in Fig. 1.10. At a fraction of a second after the removal of the partition, this

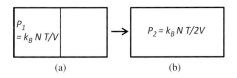

Figure 1.10 (a) N molecules of a gas, initially confined to a volume V, and (b) finally occupying the larger volume $2V$.

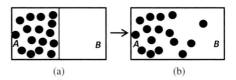

Figure 1.11 (a) A gas confined to an initial volume V. (b) A short time after the removal of the partition.

equation of state will not hold. In fact, the pressure of the system is not even defined at that moment. If we were to measure the pressure of the gas at different points in the system we would measure different pressures at different points, and those readings on our barometers will change with time. For instance, the pressure at point A in Fig. 1.11 will be almost the same as the one we measure before the removal of the partition. On the other hand, we might measure a nearly zero pressure at the point B on the far right hand side of the compartment. For a non-ideal gas, the temperature might also change when the gas expands. Thus, it is clear that right after the removal of the partition, and before a new equilibrium state is reached; the thermodynamic state of the system is not well-defined.

Experience tells us that after *some time*, the system will reach a new equilibrium state, and in particular, the pressure will be the same at each point in the system, and will not change with time. More specifically, the new state is characterized by N, T and $2V$, and the corresponding pressure is $P_2 = k_B NT/2V = P_1/2$.

We cannot tell how long it will take for the system to reach an equilibrium state. Some systems such as gas at room temperature will reach equilibrium in a few seconds. It will take a shorter time if the temperature is higher (i.e. the average kinetic energies of the particles is large). It will be slower for a liquid, and more so for solids, and it could take eons for some glasses at low temperature.

Thus, thermodynamics does not tell us anything about the *speed* or the *rate* at which the system will reach an equilibrium state. It assumes that such a state *exists* and all the relationships between the thermodynamic parameters apply to this equilibrium state. In practice, it is not easy to tell whether a system is, or is not at equilibrium.

Note carefully that the thermodynamic parameters and the relationships between them *apply* only to equilibrium states. On the other hand, equilibrium states are *defined* as those for which these relationships apply. Thus, we see that this *definition* of a thermodynamic equilibrium is circular. We can say that thermodynamic parameters and the relationships between them apply to equilibrium states, and equilibrium states are described by the thermodynamic parameters, and their relationships.
As Callen (1985) put it:

> *"In practice, the criterion for equilibrium is circular. Operationally, a system is in an equilibrium state if its properties are consistently described by Thermodynamics Theory! And Thermodynamic Theory applies to equilibrium states."*

Notwithstanding the circular definition of the equilibrium state, the existence of such states is a fundamental assumption of thermodynamics. In fact, the existence of equilibrium states is as important as the First, or the

Second Law of Thermodynamics, and certainly more important and more general than the Zeroth or the Third Law of Thermodynamics. We shall see in Chapter 4 the intimate relationship between the tendency towards equilibrium, and the Second Law of Thermodynamics. We shall also discuss a new definition of equilibrium state in terms of maximum of Shannon's Measure of Information.

There are cases when it is clear that the system is *not* in an equilibrium state, yet one can apply thermodynamics. For instance, two compartments with a very small hole between them, or two isomers with tiny amount of catalysts. In both cases, the system is not at equilibrium — e.g. its density changes with time. There is a leak from one compartment to another or from one isomer to another. Yet, the changes are so slow, and if we stop the "leak," the system will immediately be at an equilibrium state.

It should be noted that the properties of glasses can change very slowly, yet one cannot apply thermodynamic relationships to such systems. The slow processes discussed above are such that we can safely say that they proceed along a sequence of equilibrium states.

1.4 THERMODYNAMIC STATES AND STATE FUNCTIONS

An important concept in thermodynamics is the *state function*. We already introduced the concept of equilibrium and the description of the thermodynamic state by a few thermodynamics parameters. Once these parameters are specified they determine all other parameters that can be defined for the system.

We provide a simple example here to emphasize the freedom we have in choosing the *independent* parameters to describe the system and the corresponding dependent parameters which are referred to as state functions. Suppose we have an ideal, one-component gas at equilibrium. We may choose to characterize the system by specifying its temperature (T), pressure (P), and number of particles (N). Once we have made this choice, the volume (V), the energy (E), the entropy (S), etc. are said to be *state* functions. They are *determined* by the specification of (T, P, N). It is important to emphasize that these state functions have values only when the system is at *equilibrium*. For instance, the volume of the system is defined whenever (T, P, N) are fixed and do not change with time.

Consider the two compartments shown in Fig. 1.11a. Initially, all the particles are in the left compartment while the right one is empty. The initial system is characterized by (T_1, P_1, N) and the volume is V_1. Once we remove the partition there is a period of time when the system is not at equilibrium. At this point in time, the volume of the system is not well defined. It is certainly not the volume V_1. The system will have a new well-defined volume, at the new equilibrium state. The new volume is V_2, which in our case is determined by (T_1, P_2, N). Note that for an ideal gas, the temperature and the number of particles will not change in this process. If the gas is not ideal, then the temperature of the gas might change in this process.

Thus, we can say that the volume of the gas is well-defined in the initial and in the final equilibrium states. It is not defined in any intermediate state when the system is not at equilibrium. For this reason, it is meaningless to talk about the change of the volume as a function of time while the process of expansion takes place. In thermodynamics, one discusses an idealized process of expansion from an initial to a final state along a sequence of intermediate equilibrium states. We shall refer to such a process as a *quasi-static process*. It is only in such a process that it will be meaningful to talk about a (nearly) continuous change in the volume, or any other state function of the system.

1.5 THE STRUCTURE OF THERMODYNAMICS

Although we shall not discuss applications of thermodynamics in this book, it is appropriate to present here what may be referred to as the *structure* of *thermodynamics*. We have already discussed the various parameters with which we define the various thermodynamic systems. Here, we introduce the concept of the *fundamental function*, or the fundamental equation for a specified thermodynamic system. We shall see in Chapter 4 that these fundamental functions are used in distinguishing between the various formulations of the Second Law. In this section we use the term "structure" not in its usual meaning, but to convey the idea that thermodynamics deals with different systems, each described by different set of variables, and for each set of variables corresponds a different formulation of the Second Law.

As we have discussed earlier, it is an experimental fact that a system of a very large number of particles, say

$N \approx 10^{23}$ may be described thermodynamically by a small number of thermodynamic parameters.

For instance, a one-component system consisting of one phase may be described by two intensive variables, say P and T. If there are c components and p phases, the *phase rule* states that we need $f = c - p + 2$ intensive parameters, where f is referred to as the number of *degrees of freedom* of the thermodynamic system. If we want also to describe the size of the system, we need to add at least one extensive parameter, say the volume, or the energy. For instance, the equation of state of an ideal gas may be written as $P = \rho RT$. Here, R is the gas constant, P the pressure, ρ the density ($\rho = N/V$), and T the temperature. Clearly, specifying any two of these parameters (say P and T), determines the third (say, ρ). However, if we want to describe the size of the system we need to split the ratio $\rho = N/V$, into N and V, and write the equation of state as $PV = NRT$. In this description, we need three parameters; say P, T and V, or P, T, and N, to describe the system.

In thermodynamics, we have the liberty to choosing the independent variables to describe the system. From now on, we shall assume that the system consists of a one-component, and that there are no external fields. In such system, we need three parameters, say, (T, P, N), (T, V, N), (T, V, μ), etc. (μ is the chemical potential which was mentioned earlier and will be mentioned in other places in this book).

This freedom of choice of parameters is also a potential source of confusion, and ambiguity. For instance, in mathematics when we have a function, say, $z = f(x, y)$, and we talk about the partial derivative of f with respect to x, it is clear that we refer to the derivative of f with respect to x, keeping y unchanged. Qualitatively, this is a measure of how much z changes when we change only x by the amount of Δx, but keep y unchanged. Also, when we say that the function has a maximum with respect to x, it is clear that the value of z is maximum with respect to all values of x (in certain range of change of x) when y is kept constant.

The situation in thermodynamics is very different. Because of the freedom of choice of the independent parameters, we can choose either P, T, N, or E, V, N to describe the system. Once we have chosen the independent variable, we can express any other thermodynamic quantity in terms of these three chosen parameters. For instance, the entropy (which will be discussed in Chapter 4) can be viewed as either a function of P, T, N, or of E, V, N. This is usually written as $S = S(P, T, N)$ or $S = S(E, V, N)$. Note that the letter S is used here both as the *value* of the entropy, and as the *name of function* which connects between the independent variables P, T, N, or E, V, N, and the entropy. Clearly, these two are two different *functions*, and should be indicated by different letters, say $S = f_1(P, T, N)$, and $S = f_2(E, V, N)$.

Here is a potential source of ambiguity. When we say that the derivative of the entropy with respect to N is some number, it is not clear which function we use: $S = S(P, T, N)$, or $S = S(E, V, N)$. The derivatives of these two functions with respect to N are different for the different choices of the independent variables.

Writing a derivative of the form $\partial S/\partial T$ is meaningless unless one specifies the variables which are held constant in performing this derivative. Equivalently, unless we specify the independent variables, chosen to describe the system, say, (T, P, N), or (E, V, N).

We now turn to discuss the *structure* of thermodynamics and the corresponding formulations of the Second Law.

A convenient starting description of a thermodynamic system is with the variables E, V, N (one component system, having energy E, volume V, and N particles). This particular set of variables corresponds to an isolated system, i.e. when E, V, and N are fixed. In such a system, the *fundamental equation* is the *entropy function* which we write as $S = S(E, V, N)$. We can change systematically the independent variables, say from (E, V, N) to (T, V, N), to (T, P, N), etc. In thermodynamics, the transformation from one set of variables to another is achieved by the Legendre transformation. This is well described in Callen (1985). In statistical mechanics, the change of variables is achieved by the Laplace transform (or the discrete analog of the Laplace transform). This is described in most textbooks of statistical thermodynamics [e.g. Hill (1960), Ben-Naim (1992)]. Both of these transformations are mathematical. We shall later describe an equivalent "experimental" procedure of transformation from one set of variables to another. This is schematically shown in Fig. 1.12. At the top of this diagram we put the variables (E, V, N) corresponding to an isolated system. On the second row, we have changed one extensive variable to one intensive variable. On the third row we changed one more extensive variable. If we have many components we can change each of the N_i by the corresponding chemical potential μ_i.

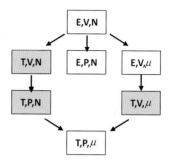

Figure 1.12 The general scheme of changing variables, starting from E, V, N and changing one variable at a time. The more common variables are shown with grey background.

For each set of independent variables, there is a fundamental equation, and the corresponding thermodynamic *potential*, or thermodynamic quantity, which is a function of these variables, and has the unique property of having a maximum, or minimum, keeping these variables fixed. This sounds a little abstract so let us discuss a specific example. An isolated system is characterized by fixed E, V, N. The entropy function is $S = S(E, V, N)$. In Chapter 4 we shall formulate the Second Law in terms of this function. It will be shown that the entropy S has a maximum. The question is, maximum with respect to what?

Again, in mathematics when we say that a function $z = f(x, y)$ has a maximum with respect to x, it is clear that z has a maximum value when varying x, but keeping y constant. The situation is very different in thermodynamics. Because of the freedom of choice of the independent variables, we could have written a few different functions for the entropy, e.g. $S(E, V, N)$, $S(E, P, N)$

or $S(E, V, \mu)$. The question is which of these has a maximum? And the next question is: maximum with respect to which variable?

Regarding the first question, only in the first function $S(E, V, N)$ (i.e. the entropy of an isolated system characterized by *fixed* values of E, V, N), does the entropy have a maximum. This is also the system for which the entropy formulation of the Second Law applies. This is further discussed in Chapter 4.

The answer to the second question is more subtle. First, contrary to the "mathematical tradition," the function $S = S(E, V, N)$, has a maximum *not* with respect to any of the independent variables E, V or N, but to another variable, while keeping E, V, N constants. The "other variable" could be an internal single parameter, a vector, or a whole function. The simplest case is the location of a movable partition between two compartments, Fig. 1.13. Starting with any location x' of the partition, after releasing the constraint on the partition, the system will move to a new value of $x^{(eq)} = x''$ for which the entropy is maximum. Thus, we can say that the function $S = S(E, V, N; x)$ has a maximum with respect to x, keeping E, V, N

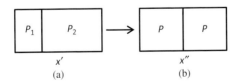

Figure 1.13 (a) A movable partition at location x', separating two compartments. (b) The equilibrium location of the partition at x'', at which the pressures in the two compartments are equal.

constants. A more complicated case is a molecule with an internal rotational angle ϕ. Here, again we can prepare, at least theoretically (or in practice we envisage an inhibitor which does not permit the change in the angle ϕ), a system with an arbitrary initial distribution of angles $x(\phi)$. When we remove the inhibitor, i.e. the constraint on a fixed distribution of angles, the entropy will increase. We can say that the function $S = S(E, V, N; x)$ has a maximum with respect to all possible *distributions* x, keeping E, V, N fixed. Note that the vector x stands for the *entire function* $x(\phi)$. This formulation of the second law applies only to an isolated system, i.e. a system characterized by the fixed values of the variables E, V and N.

For any other independent variables the formulation of the second law is different. We shall discuss some of these in Chapter 4. For instance, for a system at constant (T, P, N), the fundamental function is the Gibbs energy, given by $G = G(T, P, N)$. It is only for this function that the formulation of the second law applies: namely for a system at constant T, P, N, the Gibbs energy functional $G(T, P, N; x)$ has a *minimum* over all possible distributions x of some internal property (say, rotational angle).

Thus, we conclude that the *structure* of thermodynamics is determined by the different choices of the independent parameters which describe the system. For each set of parameters one can define the corresponding *potential function*, and the *fundamental equation*. The second law is formulated differently for each of these systems in terms of the corresponding potential function.

We shall further elaborate on the different formulations of the second law in Chapter 4.

At this point we end our introduction of thermodynamics. The next section is a brief introduction to Shannon's measure of information (SMI). The reader is urged to read carefully this section, since it is essential to understanding the concept of entropy as discussed in Chapter 4.

1.6 A BRIEF INTRODUCTION TO SHANNON'S MEASURE OF INFORMATION (SMI)

In this section I will present, in a very qualitative language, the concept of the SMI. This concept is of interest in its own right, even without any reference to thermodynamics. I believe that every curious and intelligent person should be familiar with this interesting, useful and beautiful concept. In Chapter 4 we shall see that this concept is essential to the understanding of entropy and the Second Law of Thermodynamics.

Entropy has been, and still is, one of the most misunderstood concepts in physics. It has been a mystery for a very long time, and even today many physicists consider it to be a concept that has eluded explanation for decades. In fact, one of my friends and colleagues once told me that he believes that entropy will forever be a mystery. In Chapter 4 we shall show that defining entropy in terms of SMI provides a simple and irrefutable interpretation of entropy. It also removes any traces of mystery from entropy and the Second Law.

1.6.1 The Four Steps of Obtaining Entropy from the SMI

In the following subsections, we describe the process of obtaining entropy from the SMI in four steps. First, we start by playing a simple and familiar 20-question (20Q) game. I choose an object out of n possible objects, and you have to find out which object I chose by asking binary questions, i.e. questions which are answerable by either Yes or No. We assume in this game that the probability of choosing the object out of n objects is the same for each object and is equal to $1/n$. We will refer to this game as the *uniform* probability game, or the uniform game for short.

To make the game more precise and to understand the generalization of the game, consider the following simple 20Q game.

You are shown a board divided into 10 equal regions, (Fig. 1.14a). A dart is thrown by someone blindfolded. Your task is to find in which of the 10 regions the dart is, by asking binary questions.

In the second step we generalize the simple 20Q game described above to a more complicated game where the events are chosen with different probabilities.

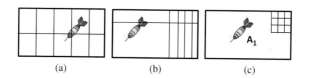

Figure 1.14 (a) A uniform, and (b) and (c) two non-uniform games, all having the same number of regions $n = 10$.

Introduction: Basic Concepts of Thermodynamics 27

To understand this more general game, consider a board at which a dart was thrown. The board is divided into 10 regions with *unequal* areas, and the dart was thrown by someone wearing a blindfold. Now you are told that the dart hit a point on the board. You are also informed about the relative areas of all the regions in Fig. 1.14b. Your task now is to find in which part of the board the dart is. To do so you can ask only binary questions. You should realize that this game is different from the previous game because the "events" here are not equally probable.

A quick question before we proceed. Suppose you are offered to play the 20Q game on either the board of Fig. 1.15a or 1.15b. You have to pay one dollar for each question you ask. Once you find out where the dart hit, you get a prize of 20 dollars. Which game would you choose to play?

Next, we turn to the third step. If you understand the 20Q game, and if you can answer the question I posed above (about the preferable game out of the two in Fig. 1.15), you should realize that in playing the 20Q game, you *lack information* on the location of the dart. By

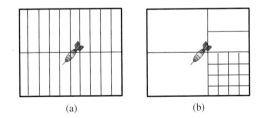

Figure 1.15 (a) A uniform, and (b) a non-uniform game with the same number of regions $n = 20$.

asking binary questions you *gain information* from each answer you receive (and you have to pay to get that information). Eventually, you will obtain all the information you need.

The third step does not introduce any new *generalization* to the 20Q game. It only makes the same game very large — very much larger than the games you are used to play, or games shown in Fig. 1.14 or 1.15. The game is large indeed, but there is nothing new in principle. You cannot possibly play this game because you do not have enough time in your life to ask so many questions — but you can at least *imagine* playing such a game. It will perhaps be called the 10^{23} Q game, rather than the 20Q game. You can be assured that if you only imagine playing this game, you will be effortlessly led to the fourth and last step; the understanding of the concept which is still considered by many as the most mysterious concept in physics — the entropy. Let us proceed with the more quantitative aspect of the 20Q game

1.6.2 First Step: Playing the Uniform 20Q Game

We start with a relatively easy game, like that of Fig. 1.15a. A dart hit a board which is divided into n equal-area regions. You are told that the dart was thrown by someone blindfolded, and it hit one regions of the board. You also know that the probability of hitting any one of the n regions is $1/n$. This is why we refer to this game as the *uniform* game.

For = 8, how many questions do you need in order to find out where the dart is? If you are smart enough, you can obtain the information on where the dart is in three questions. How?

You simply divide the 8 regions into two groups, and ask, "Is the dart on the right group?" If the answer is *yes*, you divide the remaining four regions on the right into two groups, and ask, "Is it the right group," and so on. Thus, each question eliminates one half of the remaining regions. In this method you will find out where the dart is in exactly three questions. This is referred to as the smartest strategy, (Fig. 1.16a). One can prove mathematically that by asking questions of this kind (i.e. dividing each time into two equally probable regions) you get *maximum information* for each answer. Therefore, you will need to ask minimum number of questions to get the information you need. It is interesting to note that young children

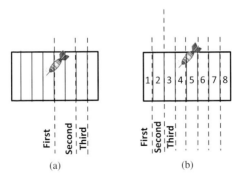

Figure 1.16 Two strategies for asking questions. The dashed lines shows the division of the total number of possibilities into two parts; (a) the smartest, and (b) the dumbest strategy.

tend to ask *specific questions*; is the dart in region 1, or is it in region 2, etc.? This is referred to as the "dumbest" strategy, (Fig. 1.16b). They do not have patience to plan the "strategy" of asking questions. At about age 14–15, children realize that it pays to choose the smartest strategy. Although there is zero chance of winning the game on the first few questions, the smartest strategy is more efficient on average (for more details on this see Ben-Naim (2010)). Figure 1.17 shows the average number of questions, as a function of n for the "smartest" and the "dumbest" strategies. For more details, see Ben-Naim (2008).

Take note that the number of questions is related to the number of regions by the logarithm to the base 2. Thus, for $n = 8$, we have $3 = \log_2 8$. You can check for

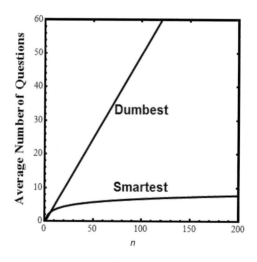

Figure 1.17 The average number of questions as a function of the number of objects n, in a 20Q game, for the "dumbest" and the "smartest" strategies.

yourself that for $n = 16$ regions you will need four questions. For $n = 32$, you will need five questions, and for $n = 2^k$ with an integer k, you will need to ask k questions. Take note that when we *double* the number of regions, the number of questions you need to ask (smart questions) *increases* by *one*. It can be proved in general that for any integer n, the average number of (smart) questions you need to ask is approximately $\log_2 n$. We say "on average" because for any n, say $n = 7$, you cannot divide at each step the total number of regions into two equally probable groups. But you can try to do it as closely as possible: for instance, divide the 7 regions into 4 and 3 regions. The general result for any number n of equally probable regions is that you need to ask, on the average about $\log_2 n$ questions in order to obtain the required information. All this is valid for a system of *equally probable regions*. It is very easy to prove that the relationship between the number of regions and the number of (smart) questions you need to ask depends on n as $\log_2 n$. The reader is urged to try to play the 20Q game with different numbers of regions n. A simulated game of this kind may be found in ariehbennaim.com [→ books → entropy demystified → simulated games].

1.6.3 Second Step: Playing the Non-uniform 20Q Game

By non-uniform game, we mean unequal probabilities for the different events. To be specific, consider the modified game shown in Fig. 1.14b. Here, we divide the board into 10 regions. Unlike the game in Fig. 1.14a, here the areas are different, and hence the probability of hitting each area is different.

Usually, when we play the parlor game of 20Q game we assume implicitly that the probability of choosing a specific object or a specific person is $1/n$, where n is the total number of objects from which one is selected.

In the example shown in Fig. 1.14b, the areas of the regions are not equal. This makes the *calculation* of the number of questions more *difficult*, yet the actual *playing* of this game is *easier* than the one shown in Fig. 1.14a. If you are not convinced that the game in Fig. 1.14b is easier to play, consider the more extreme game in Fig. 1.14c.

One can prove mathematically that the average number of questions you need to ask in order to obtain the information (on where the dart is) is given by the famous Shannon formula

$$H = -\sum p_i \log_2 p_i$$

where p_i is the probability of the event i, which in our case is simply the relative area of the region i. We shall refer to H as the Shannon Measure of Information (SMI). Figure 1.18 shows the function H for the case of two outcomes with distribution $(p, 1-p)$.

This is a remarkable relationship. Having a 20Q game with any given distribution of the objects (or the outcomes, in a general experiment), it can also be proved that the quantity H defined above is *always* smaller than the quantity H defined on a *uniform distribution* with the same total number n. In terms of 20Q game we can say that playing the game of Fig. 1.14b is always *easier* than

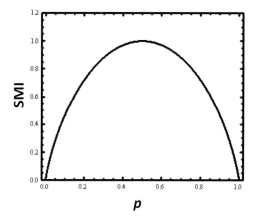

Figure 1.18 The Shannon Measure of Information for the case of two outcomes.

playing the game in Fig. 1.14a. Easier in the sense that you will need, on average fewer questions to ask in order to obtain the information you need. I hope you will also realize that the game in Fig. 1.14c is the easiest of the three games shown in this figure.

This mathematical result is quite obvious intuitively. If you are offered the two games in Fig. 1.14a and 1.14b, and you have to pay one dollar for each answer, and get a prize x when you get the information, it is always advantageous to play the *non-uniform* game. You can try playing the two games and convince yourself that the uniform game always requires, on average more questions. You may also try to play the more extreme game shown in Fig. 1.14c. Here, if you know the distribution, you should ask the first question: "Is the dart in area A_1?" The chances to get a Yes answer are about 9/10. This means that if you play the game many times, you will get the information (on where the dart is), in about one question. Shannon's Measure of Information provides the general relationship between the number of (smart) questions to ask, and the probability distribution of the events.

1.6.4 The Third Step: Generalization from a 20Q Game to an Over 20^{23}Q Game

Now that you have an idea about the relationship between the number of objects, and the average number of binary questions you need to ask in order to find out which one of the objects was selected, let us play a very large game of the same kind.

Suppose you have a single atom in a cubic box of edge d, (Fig. 1.19). The *state* of the atom can be described by its location and its velocity at each instance of time. Let l be its location and v its velocity. Clearly, the pair (l, v) describes the *state* of this atom. It is also clear that there is an infinite number of states. Therefore, if I know the state of the atom, and you have to find its location by

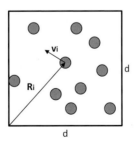

Figure 1.19 Particles in a box of volume V. Each particle i has a locational (**Ri**) and a velocity (**Vi**) vector.

asking binary questions, you will need on average, to ask an *infinite* number of questions.

Fortunately, there is the uncertainty principle in physics. This principle states that you cannot determine both the exact location and velocity of the atom, but there is a limit to the "size" of the box" ($\Delta l \Delta v$) in which you can determine the state of the atom. This passage from the infinite number of states in the *continuous* range of locations and velocities, to the finite number of possibilities is shown schematically in Fig. 1.20.

Now, if I know the state of the atom and you have to ask binary questions, you will need to ask only a finite number of questions. This game is no different from playing the 20Q game that you are familiar with. Next, we move from one atom to a huge number of atoms, say 10^{23} atoms in the same box of edge d. The problem is now to find the "state" of this huge number of atoms, each being in a tiny box of size ($\Delta l \Delta v$). Again, there is no difficulty in playing this game. You will need to ask many

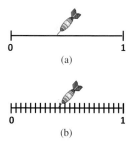

Figure 1.20 Passage from the infinite (continuum) to the finite (discrete) description of the states in one-dimensional systems.

questions, far too many that you can achieve in your lifetime, or during the whole age of the universe. However, there is no difficulty in principle in *imagining* playing such a huge game. There will be a finite number of questions — finite, albeit a huge number. Thanks to Shannon's measure, it is possible for us to actually calculate the average number of questions we need to ask ("smart questions") in order to find out the microscopic state of any number of particles. All of the above discussions refer to an ideal gas, i.e. a system of non-interacting particles. For a real gas or a liquid, we need to also take into account the interactions among the particles. This does not concern us here [see Ben-Naim (2008)].

Finally, we need to introduce one principle from physics to reach for the entropy; the particles are indistinguishable. This means that if you exchange the locations of two atoms, you get the same configuration. Figure 1.21 illustrates this reduction in the number of configurations for three particles.

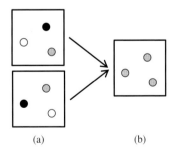

Figure 1.21 (a) Two different configurations of three distinguishable particles become identical, (b), when the particles are indistinguishable.

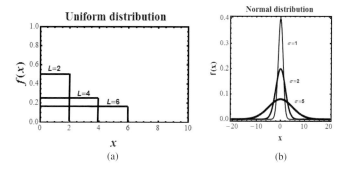

Figure 1.22 (a) Uniform and, the (b) normal or Maxwell-Boltzmann distribution.

Remember that the SMI has a single maximum at some specific probability distribution. For our system we have to assume that the system is at equilibrium. For an ideal gas this means that the probability of finding a particle in a given interval Δx is independent of x. The same is true for the y and z axis. Figure 1.22a shows the uniform distribution in one dimensional system.

Similarly, the equilibrium velocity distribution is the one shown in Fig. 1.22b (here we show the probability density for different values of the standard deviation σ). It is called the Maxwell-Boltzmann distribution. This is the distribution of velocities that maximizes the SMI. For more details, see Ben-Naim (2008, 2012, 2014, 2015a).

1.6.5 The Final Step: The Entropy Spawns as a Particular Case of a 20Q Game

Now we reach the most exciting point. We have to do two steps to get the *entropy* out of the SMI:

First, we need to apply the SMI to the distribution of locations and velocities of all the particles in the system.

Second, we have to apply the SMI for the specific distribution of locations and velocities at *equilibrium*.

Once you do this you get a quantity which, up to a multiplicative constant, is identical with the *thermodynamic entropy*. We shall further discuss the thermodynamic entropy in Chapter 4. For now the reader is urged to remember that we give the name *entropy* to the specific value of the SMI defined on the distribution of locations and velocities at equilibrium.

Incredible as it may sound, entropy, a term which came into being in the field of science due to the study of the efficiency of heat engines, is nothing but the average number of questions one needs to ask in order to find out which microscopic state the system is in, presuming that the system is in a well-defined macroscopic state at equilibrium.

1.7 SOME CONCLUDING REMARKS

In this chapter, we did not introduce any of the laws of thermodynamics. The most important concept introduced in this chapter is the concept of *thermodynamic equilibrium*. It has the same flavor as the concept of equilibrium in mechanics; it is stable in the sense that it does not change with time. If there is a small fluctuation in the state of the system, then there will be a "restoring force" which will bring back the system to equilibrium. Thermodynamics deals with equilibrium states. All the parameters describing the system, as well as the state functions apply to equilibrium states. We also introduced the concept of SMI. This is an important quantity with various applications in many fields of science. Here, I claim that this quantity is indispensable for understanding the meaning of entropy and the Second Law of Thermodynamics.

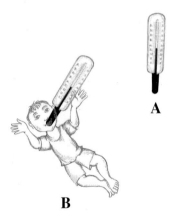

A

B

C

2

The Zeroth Law (ZL)

The Zeroth Law of Thermodynamics was first suggested by Ralph Fowler. It got the name "Zeroth Law" simply because the first and the second laws had already been formulated. The Zeroth Law (ZL) can hardly be reckoned as a law in thermodynamics, or in any other field of science. It is a statement about the property of the thermal equilibrium, but the same property also applies to mechanical and material equilibria. In Note 1 we present a few quotations from Atkins' book (T4L) on the ZL.

2.1 THE THERMAL ZEROTH LAW

Typically, the Zeroth Law (ZL) of Thermodynamics is formulated as follows:

If two systems, A and B are in thermal equilibrium with a third system, C, then A and B must also be in thermal equilibrium with each other. Clearly, since this Law deals with *thermal* equilibrium it is more appropriate to refer to it as the *Thermal-Zeroth-Law*. Assuming that we intuitively know what "thermal equilibrium" means, this law seems to be almost self-evident.

In some textbooks one might find a formulation of the ZL as follows:

Thermal equilibrium is a transitive property.

A *transitive* relation between two objects is defined as follows[2]: If A is related to B, and B is related to C, then A is also related to C. Denote the relation *thermal equilibrium* by TE, the transitivity of thermal equilibrium can be formally written as (see Fig. 2.2):

Transitivity of TE:

Given: A is in TE with B
and: B is in TE with C
then: A is in TE with C

Note however that from the definition of transitivity (Fig. 2.1), as applied to the TE, the ZL, as formulated above (Fig. 2.2) does not follow, i.e.

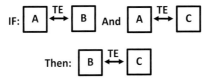

Figure 2.1 The Zeroth Law.

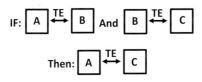

Figure 2.2 Transitivity of Thermal Equilibrium (TE).

The ZL:

Given: A is in TE with B
and: A is in TE with C
then: B is in TE with C

Clearly the ZL does not follow from the transitivity of the TE. The ZL follows from the transitivity of the TE only when one adds the symmetric property of the TE. In fact the relation TE is actually an *equivalence* relation. An equivalence relation R is defined by three properties:

1. Reflexive: $a R a$
2. Symmetric: $a R b$ implies $b R a$
3. Transitive: $a R b$ and $b R c$ imply $a R c$

Clearly, TE is an equivalence relation. It is easy to see that from the *symmetric* and *transitive* properties of the TE, the formulation of the ZL (as illustrated in Fig. 2.1) follows.[1]

Now that we have an intuitive idea about the ZL, let us examine our understanding of this law.

How do we know when "A is in thermal equilibrium with B?"

We certainly cannot *see* or *feel* when two bodies are in thermal equilibrium. The way we can tell when two systems are in equilibrium is by measuring the temperature of A and B. If we bring A in thermal contact (i.e. through heat conducting wall — referred to as *diathermal*) with B, while measuring the temperatures of A and B, and if we notice no change in the temperatures of A and B, then we conclude that no heat had flowed from A to B, or from B to A. In other words, equality of temperatures is a necessary condition for thermal equilibrium.

For this reason one can reformulate the ZL in terms of temperatures.

If the temperature of A is the same as that of B, and the temperature of A is the same as that of C, then we can conclude that the temperature of B is also the same as that of C.

In some textbooks, you might read statements such as: The ZL *introduced* the concept of temperature. This is of course, not true. The concept of temperature, as well as the measurement of temperature was introduced long before the formulation of the ZL. Perhaps a more sensible statement would be that the ZL is the basis on which measurement of temperature was established.

If A is a device to which we refer to as a *thermometer* is brought into contact with B and C, and we read the same number on the scale of A (which actually is a reading of some property which changes with temperature, like the length of a metal bar, or the volume of liquid mercury in a thin capillary), we conclude that B and C have the same temperature as that of A, and therefore are at the same temperature.

At this point we must emphasize that the measuring device we use in measuring the temperature must be small compared with the body, the temperature of which is being measured. This is important, because otherwise the thermometer itself might affect the temperature of the body it measures. When we measure our body temperature (which is presumed to be constant, to a good approximation throughout the whole body) we use a thermometer which is so small that its temperature after reaching equilibrium with our body is equal to the body's temperature. This is possible only if the size of the thermometer is smaller than the size of the body. Suppose that initially, the temperature of the body is T_B, and the temperature of the thermometer is nearly the room temperature T_R. After reaching thermal equilibrium, the temperature of the thermometer will almost be T_B, and the temperature of the body almost unchanged.

Think of what will happen if you try to measure the temperature of a small insect with the same thermometer. Suppose that we have two insects that are initially at temperatures T_1 and T_2, (Fig. 2.3). The thermometer which is much larger than the insects is initially at temperature T_R. Clearly, after the "measurement" of the temperature by this thermometer, the two insects will be at thermal equilibrium!

46 *The Four Laws That Do Not Drive The Universe*

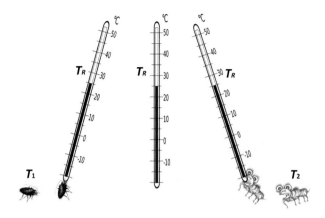

Figure 2.3 Two insects initially at different temperatures T_1 and T_2. The thermometer is much larger than the insect., and its temperature will not be affected by thermal contact with the two insects.

The "size" comment is also relevant for the formulation of the ZL. Suppose that A is very large compared with both B and C. In such a case, when we bring A into thermal contact with B and with C, it will "impose" its temperature on both B and C. In this case, B and C will be in thermal equilibrium *after* being in contact with A, but may not have been in thermal equilibrium *before* being in contact with A.

So far we have discussed the meaning of *thermal equilibrium*, and how it is determined by measuring the temperature. We assumed that we know the meaning of equilibrium, see Section 1.3. In the next two sections we shall discuss two *additional* Zeroth Laws, which in my opinion are equally important as the *thermal* ZL discussed above.

As for the importance of ZL, it should be said that many textbooks on thermodynamics do not even bother to mention this law. In my opinion, this property of thermal equilibrium does not deserve to be called a law, and certainly its importance cannot be compared with either the First, or the Second Law of Thermodynamics. Besides, the fact that at thermal contact the two bodies must have the same temperatures can be deduced from the Second Law.

In the next two sections, we formulate a Zeroth Law for *mechanical equilibrium*, and a Zeroth Law for *material equilibrium*. We briefly discuss these two here although they should be discussed within the Second Law (Chapter 4).

2.2 THE MECHANICAL ZEROTH LAW

For mechanical equilibrium we can state the corresponding Zeroth Law as follows:

If A is in mechanical equilibrium with B and with C, then also B and C are in mechanical equilibrium.

In the context of thermodynamics, two systems are said to be at mechanical equilibrium when they are in contact through a movable wall or a piston, and the wall does not move. If the pressure on one side of the wall is larger than the pressure on the other side, then the wall will move until a *mechanical* equilibrium is established between two systems. At that point the pressure on both sides of the wall will be equal. Clearly, this law belongs to mechanics. Its derivation from thermodynamics is somewhat subtle, for details see Callen (1985) Appendix C.

As in the case of thermal equilibrium the fact that mechanical equilibrium is transitive, is essential for measurement of pressure. If we can bring a measuring device A (barometer) into mechanical contact with two systems B and C, and we find the same reading of the pressure on A, we can conclude that the pressure in B is equal to the pressure in C.

Again, we emphasize that the measuring device A must be small compared with the bodies, the pressure of which it is supposed to measure. It should be small enough such that when bringing into mechanical contact the device A with the body B, the pressure of B will hardly be affected by the contact with A.

2.3 THE MATERIAL ZEROTH LAW

The thermal Zeroth Law was formulated as a law in thermodynamics because temperature and thermal equilibrium were not part of physics. These concepts had to be accommodated within thermodynamics.

The analogy between the mechanical equilibrium between two particles, and two macroscopic systems, involves the concept of *pressure*, which is well-known in physics. The pressure is defined in thermodynamics exactly the same way it is defined in physics (i.e. force per unit area).

The material Zeroth Law was never formulated as a Law, neither in physics, nor in thermodynamics. Perhaps, it is too trivial for the case of a one component system, but it is far from trivial in the case of multi-component systems.

Let us first discuss the one component system. We can state the material Zeroth Law for one component systems as follows:

If A is in material equilibrium with B and C, then B and C will also be in material equilibrium.

Material equilibrium means that if A and B are in contact in such a way that a compound, say pure water, *can* flow between the two (there is material contact), then there will be no net flow of water in either direction from A to B, or from B to A.

In this case, the Zeroth Law seems trivial and seems to follow from plain common sense. The quantity which is analogous to temperature and pressure in the previous cases is the *chemical potential* of the water. It sounds obvious that if the density of water (or any other pure substance) in A is larger than in B, then water will flow from A to B.

Note that if A and B are pure substances, then if $T_A = T_B$ and $P_A = P_B$ then it also follows that the densities must be the same. This is true for any pure substance.

Thus, if A and B are pure and they are at both thermal equilibrium and mechanical equilibrium, it follows that they must also be in material equilibrium. Therefore, there is no need to formulate a material Zeroth Law if we already have a thermal and mechanical Zeroth Law. This conclusion is correct when we deal with *pure one component* systems. It is not true for two or more component systems. For instance suppose that A and B are two systems, each containing a mixture of water (w), and ethanol (e). For simplicity, we assume that we have a one phase mixture of w and e. In this case, if A and B are in both thermal and mechanical equilibrium ($T_A = T_B$ and $P_A = P_B$)

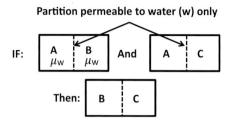

Figure 2.4 Two compartments containing water and ethanol μ_W is the chemical potential of water. If A is in water-equilibrium with B, and A is in water-equilibrium with C, then B is in water-equilibrium with C.

then it does not follow that the two systems are in either water-equilibrium, or in ethanol equilibrium.

Therefore, for such cases we need to formulate a material Zeroth Law for each compound. For instance, if A and B are brought to material contact with respect to water, i.e. the two systems are in contact through a wall (or a membrane) which is penetrable to water but not to ethanol. We shall say that A and B are in *water equilibrium*, (Fig. 2.4). For such a water-equilibrium we can formulate the corresponding Zeroth Law as follows: If A is in water-equilibrium with B and with C, then B and C are also in water-equilibrium. Similarly, we can formulate the Zeroth Law with respect to ethanol equilibrium.

2.4 THE LAW OF EXISTENCE OF EQUILIBRIUM STATES OF A THERMODYNAMIC SYSTEM

In the previous sections we discussed three different Zeroth Laws of thermodynamics. In fact, we could have formulated many more ZLs; one for each compound.

In all of the discussions above, it is presumed that the meaning of the concept of equilibrium is known. Indeed, we have an intuitive sense that we know what thermodynamic equilibrium means. However, a precise definition of this state is elusive. Here, we discuss the equilibrium state as we understand it intuitively. In Chapter 4, we shall discuss the concept of equilibrium state, or states as defined microscopically by the probability distributions of the locations and velocities of the particles.

We know from daily experience that any isolated system will evolve towards a state which is characterized by a small number of thermodynamic parameters. We refer to this state as an equilibrium state. By definition, once this state is reached, the thermodynamic parameters which characterize this state are time independent. This statement is also valid when we examine two or more systems at thermal, mechanical and material equilibrium.

Thus, the existence of an equilibrium state of a given thermodynamic system may be viewed either as a basic postulate of thermodynamics, or as a fundamental law of thermodynamics. This is far more fundamental than the ZL as formulated for thermal equilibrium. Needless to add that, however one chooses to formulate the ZL, it is clear that it *drives* no process, and certainly not the entire universe!

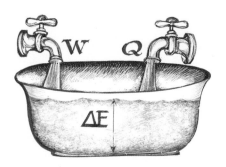

An Analog of the First Law

3

The First Law (FL)

The First Law (FL) is essentially an extension of the law of conservation of energy which was formulated in physics long before the development of thermodynamics. In this sense it is conceived as "simple," "natural" and easy to comprehend. To the best of my knowledge it has never raised any sense of mystery. This is in sharp contrast to the Second Law which is considered to be difficult to comprehend, oftentimes misunderstood, and considered to be a mystery. We shall discuss this in the next chapter.

The concept of energy is considered as a primitive concept, equating it with the work we do with our muscles and brawn. One can start with *force* as a primitive concept and define energy as force × length. Alternatively, we can start with *energy* as a primitive concept and define *force* as the gradient of energy. In any case this concept is considered to be "understood," at least intuitively as something equivalent with the "work" we do by using our muscles. We add here a quotation from Atkins' book on the FL.

"Moreover, like the zeroth law, which provided an impetus for the introduction of the property 'temperature' and its clarification, the first law motivates the introduction and helps to clarify the meaning of the elusive concept of 'energy.'"

As we have noted in the beginning of Chapter 2, the ZL did not introduce nor clarify temperature. Similarly, the FL did not motivate nor clarify the meaning of energy.

3.1 CONSERVATION OF ENERGY IN SIMPLE PHYSICAL SYSTEMS

Perhaps the simplest example for which the conservation of energy can be illustrated is by the collision of two hard particles, like marbles or billiard balls. A collision between two hard particles is shown in Fig. 3.1. The two spheres approach each other along the line connecting their centers, collide elastically, which means all the

Figure 3.1 Two hard spheres before and after the collision will have the same total energy.

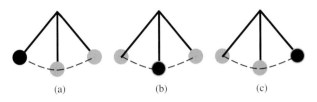

Figure 3.2 The three states of a pendulum; (a) maximum potential energy, (b) maximum kinetic energy, and (c) maximum potential energy.

kinetic energy is conserved.[3] In this process the *total kinetic energy* is unchanged in the collision.

The second example is the ideal pendulum shown in Fig. 3.2. Here, the weight W swings from left to right. In the absence of air and any other frictions, one can say that during the periodical process, the kinetic energy, which is a maximum at the bottom (Fig. 3.2b) converts to a potential energy when the weight is at its maximum height, either Fig. 3.2a or Fig. 3.2c. At all states the sum of the kinetic and potential energy of the weight is constant.[4]

There are many other process involving kinetic, electric, magnetic, etc. energies, for which the conservation of the total energy was formulated long before the formulation of the First Law (FL) of Thermodynamics.

3.2 THE EXTENSION OF THE LAW OF CONSERVATION OF ENERGY TO INCLUDE THERMAL ENERGY

In this section we introduce the First Law in two steps: We first introduce the *internal energy* for an adiabatic system. We shall discuss one kind of work, referred to as expansion (or compression) work, or *PV* work.

We assume that force is a primitive concept. Qualitatively, it is that factor which causes a change in the motion of a body. Newton's Second Law gives the relationship between the force (F) acting on a body and the acceleration (a) it imparts on the body: $F = ma$, where m is the mass of the body.

The pressure is defined as the force acting per unit area. We now apply the concept of work as known in mechanics to a thermodynamic system at equilibrium.

Consider a system in a cylinder sealed with a movable piston, (Fig. 3.3).

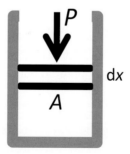

Figure 3.3 The mechanical work done on the system is equal to the pressure (P), times the area (A), times the displacement of the piston dx.

The First Law (FL) 57

The mechanical work done *on* the system is the product of the force F exerted on the piston, and the displacement dx of the piston, i.e. $F \times dx$. If the area of the piston is A then the *work* done on the system is $\delta W = F dx = PA dx = P dV$. Recall that P is the pressure exerted on the piston, and dV is the change in the volume of the system. We wrote δW rather than dW, to denote a *small* amount of work. Usually "d" is used for a differential of a state function. More on this notation will be clarified below.[5,6]

Suppose we have a system, say a gas or liquid with a *fixed* amount of matter. Such a system is referred to as a closed system, the state of which can be characterized by a few parameters, say P and T.

It is an experimental fact that the system's *state* might change when it is placed in surroundings of different temperatures. The rate of change depends on the type of material of the walls surrounding the cylinder. It is also known that some materials, which we call *thermal insulators*, preclude the change of the state of the system, when no work is exchanged between the system and its' surroundings. Later, we shall see that such insulators prevent the exchange of *heat* between the system and its surroundings. At this stage we still do not have the concept of heat or thermal energy. All we can say is that when the system is within some kind of walls (later called athermal), the change in the *state* of the system, say, from state 1 to state 2 requires a fixed amount of work.

We call a wall that is impermeable to heat flow an *adiabatic* or an *athermal* wall.

It is also an experimental fact that one can change the *state* of the system enclosed in *adiabatic* walls, say from

P_1T_1 to P_2T_2 by different kinds of work, say mechanical, electrical, etc. It is found that the amount of work required to change the state of the system enclosed in adiabatic walls, from an initial state P_1T_1 to a final state P_2T_2 is *independent* of the type of the work done on the system, or on the path along which the system's state is changed from the initial to the final state.

This fact leads to the definition of the *internal energy* of a system, denoted E, as a *state function*, i.e., a function which depends only on the *state* of the system, and not on the way that state is reached. We write this experimental finding as

$$W_{ad} = E_2 - E_1 = \Delta E$$

where E_2 and E_1 are the values of the internal energy of the system in states 2 and 1, respectively, and W_{ad} is the adiabatic work. In chemistry, the work done *on* the system is defined as *negative* and work done *by* the system *on* the environment as *positive*. In physics, the convention is different. (See notes 5 and 6.)

Thus, the internal energy, E as described above is a *state function*. It is defined up to an arbitrarily chosen additive constant. In thermodynamics, we are always interested in differences in the internal energies between two equilibrium states. Therefore, the additive constant is insignificant and therefore is not specified.

Introducing Heat Flow Into or from The System

When the walls of the system are not athermal, we find that the work required to change the state of the system from the initial to the final state might depend on the system's surroundings. The intuitive reason for that is that when the system is adiabatic, the work done on, or by the system is the only energy that is exchanged between the system and its surrounding, and therefore the change in the internal energy is exactly equal to W_{ad}. When the walls of the system are not adiabatic, allowing the flow of other kind of energy into the system (other than previously known works, such as mechanical, electrical, etc.), we assume that another form of energy flows into the system through the walls. This form of energy we call *heat*, or thermal energy. We defined the heat exchanged between the system and its environment by

$$Q = \Delta E - W$$

Note carefully the order of introducing the three quantities in the above equation. First, we start with the notion of *work* with which we are familiar from physics. Second, we recognize the existence of adiabatic walls, which we intuitively understand as being impermeable to some kind of energy that flows through the walls. We then find experimentally that if we have a system characterized by an initial state denoted 1, and we transfer a given amount of work, under adiabatic conditions, the final state of the system denoted 2, is *independent* of the type of *work* we have done, or of the path leading from the initial to the final state. It depends only on the *amount* of work, which we have denoted W_{ad}. These experiments led us to introduce the *state function E* which is defined for any *state* of the function.

Finally, when the system is not adiabatic, and when we transfer the amount of work W, we find that the system's state has changed from state 1 to state 2, but the equality $\Delta E = W$ does not hold. We suspect intuitively that another form of energy has been transferred through the walls. We refer to this form of energy as *heat*, and define the amount of heat transferred in this process by

$$Q = \Delta E - W$$

It should be clear by now that whereas E is a *state function* and is defined for any state of the system, the quantities Q and W are not state functions. Both of these are forms of energy that are exchanged between the system and its environment.

Once we have transferred either work or heat into the system, they are "stored" in the system as internal energy. There is no such thing as the "work of the system" or the "heat *of* the system", but only work done by or on the system and heat transferred into or out of the system.

To conclude, the First Law of Thermodynamics is simply an extension of the Law of conservation of energy that we already know from physics; to include also another form of energy we call heat.

To understand the difference between Q and W on one hand, and the internal energy on the other, consider a water bath that can be filled by two taps. See figures on page 52 facing the beginning of this chapter.

Through one tap, blue colored water flows into the tub, while red colored water flows through the other tap. Once the water enters into the bath the colored molecules react with each other, so that the resulting compound is colorless. We can measure the amount of water that was transferred to the bath from one tap, which we denote by Q. Similarly, we denote by W the amount of water flowed into the bath through the second tap. Once the water is in the bath we cannot distinguish between the Q-water, and the W-water. All we can measure is the total amount of water that was transferred by the taps into the bath, which we denote by ΔE. Thus, if Q moles of water were exchanged between the red tap and the bath ($Q>0$ correspond to flow into the bath and $Q<0$ out from the bath), and if W moles of water were exchanged between the blue tap and the bath ($W>0$, into and $W<0$ out from the bath), the total change in the amount of water in the bath is

$$\Delta E = Q + W$$

which, if you like, you can call it the law of conservation of water molecules.

3.3 THE END OF THE TWO SEPARATE LAWS; THE CONSERVATION OF ENERGY, AND THE CONSERVATION OF MATTER

Once the FL was formulated it seemed to be absolute and indisputable. Parallel to the conservation of energy, the law of conservation of matter was developed. Chemical reactions can transform one compound to another, but the total mass of the system is unchanged in the reaction. For instance, when a candle burns, it seems as if some matter "disappears." However, when all the gases were collected and analyzed, it was found that no matter was "lost," but transformed from one kind to another.

Once it was recognized that the burning is nothing but the transformation from one set of compounds into another set of compounds, scientists declared this fact as the law of conservation of matter. Thus, we had two separate and seemingly unrelated conservation laws: The conservation of matter and the conservation of energy. This state of the two laws was maintained even after it was shown that nuclear reactions can transform from one *element* to another, so that the total numbers of atoms of a specific element is not conserved, but the total *mass* of all atoms is conserved.

This situation was radically changed upon the discovery that in some nuclear reactions, some reduction of the mass of the nucleus was detected. This led to the law of conversion between the mass m and the energy E:

$$E = mc^2$$

where c is the speed of light.

The fact that mass can be converted to energy saw the demise of the law of the conservation of mass. The fact that energy can be converted to mass spelled the collapse of the conservation of energy. However, the end of the two laws brought forth an offspring, coalesced into one law, the conservation of *mass-energy*. This is neither the conservation of energy, nor the conservation of matter. It is an entirely new law. While the separate laws of conservation of energy and conservation of matter were accepted as "natural," the convergence of the two laws was far from being "natural." However, for many processes carried out in both physics and chemistry, one can still use the two distinct laws of conservation of matter and energy.

3.4 APPLICATIONS OF THE FL

There are very few applications of the FL. Most applications of the FL are realized when it is combined with the Second Law.

The most important application of the FL in chemistry is the so-called Hess' Law, sometimes referred to as the *additivity* of chemical reactions. In most textbooks Hess' Law is formulated in terms of enthalpy. Here, for simplicity we discuss this law in terms of energy.

Consider the following chemical reactions:

$$4NH_3 \rightarrow 2N_2 + 6H_2 \qquad (1)$$

$$2N_2 + 2O_2 \rightarrow 4NO \qquad (2)$$

$$6H_2 + 3O_2 \rightarrow 6H_2O \qquad (3)$$

$$4NH_3 + 5O_2 \rightarrow 4NO + 6H_2O \qquad (4)$$

Note that reaction (4) is a sum of the three reactions (1) + (2) + (3). We denote by $\Delta E^f(\alpha)$ the *energy* of *formation* of the compound α. This is the energy associated with the *formation* of one mole of a compound α from its elements, all at some specified standard states.

The *standard energy* change for reaction (4) may be written as:

$$\Delta E°(4) = 6\Delta E^f(H_2O) + 4\Delta E^f(NO) - 4\Delta E^f(NH_3)$$

Note that we put a minus sign before the formation energy of a reactant, here, NH_3. Also, the formation energy of the oxygen does not appear in this equation.

Hess Law states that $\Delta E°(4)$ is the sum of the standard energies of the reactions (1), (2), and (3).

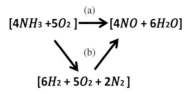

Figure 3.4 The reaction (4) may be viewed as a transformation between two states of the system. This transformation can be carried out along two paths; (a) and (b).

$$\Delta E°(4) = \Delta E°(1) + \Delta E°(2) + \Delta E°(3)$$

In Fig. 3.4, we show graphically that reaction (4) may be viewed as a transformation from "state" [$4NH_3 + 5O_2$] to the "state" [$4NO + 6H_2O$]. This transformation can be done along two paths: One, as shown as in reaction (4), and the second, as first dissociating NH_3 into its elements (N_2 and H_2), then combine N_2 and O_2 into NO and H_2 and O_2 into water. If the two paths are carried out at the same temperature and pressure, then the energy change along the two paths must be the same. Thus, the Hess Law is essentially a statement about the energy being a *state function*. The same reasoning applies to any other state function such as enthalpy, entropy, Gibbs energy, etc.

3.5 SOME FINAL THOUGHTS ON THE FIRST LAW

The law of conservation of energy is sometimes stated as a law that forbids the formation of perpetual-motion machines, i.e., the much sought after machines which are capable of doing unlimited work with a given amount of energy.

The energy drives many processes, but perhaps not all processes. Perhaps, some of our mental activities are driven by other, yet undiscovered "driving forces." But it is safe to say that energy is the main "driving force" for many physical processes. Note however, that the *First Law* does not drive anything, certainly not the entire universe. See also Chapter 6.

Finally, it is worth mentioning a beautiful, and to some extent, surprising and unexpected theorem known as "Noether's theorem." It is not a *useful* theorem; one can apply the FL without ever mentioning this theorem. Most textbooks on thermodynamics do not mention the Noether theorem. This includes the first edition of Callen's book (1961), but it is mentioned in the second edition, Callen (1985).

The theorem asserts that every continuous symmetry of the dynamical behavior of a system implies a conservation law for that system. Specifically, the conservation of energy is associated with time-translation symmetry. This means that dynamic laws of physics, such as Newton's Law, are unchanged by the transformation $t \rightarrow t' + t_0$ (i.e. by a shift of the origin of the scale of time). This beautiful theorem has been transformed, or rather deformed, by Atkins (2007) into an ugly and a meaningless theorem. On page 45 of the "Four Laws," we find:

> *Energy is conserved because time is uniform: time flows steadily; it does not bunch up and run faster then spread out and run slowly. Time is a uniformly*

> *structured coordinate. If time were to bunch up and spread out, energy would not be conserved. Thus, the first law of thermodynamics is based on a very deep aspect of our universe and the early thermodynamicists were unwittingly probing its shape.*

Symmetry under time translation has nothing to do with "time's flow," slowly or fast. It is not clear that time flows [See Ben-Naim (2016)]. I have no idea what "uniformly structured coordinate" means. I also do not have any idea what is meant by the "shape" of the universe that early thermodynamicists were, wittingly or unwittingly probing. These are beautiful words which are devoid of any meaning. Certainly they have nothing to do with Noether's theorem.

$$\Delta S = k \, Log(Pr_2/Pr_1)$$
$$\Delta A = -kT \, Log(Pr_2/Pr_1)$$
$$\Delta G = -kT \, Log(Pr_2/Pr_1)$$

Various formulations of the Second Law:

<u>Probabilistic:</u> In a spontaneous process occurring in a well defined system, the system will evolve from a state of lower probability to a state of higher probability.

<u>Thermodynamic:</u> As a result of the process, the entropy increases in an (E, V, N) system, the Helmholtz energy decreases in a (T, V, N) system, and the Gibbs energy decreases in a (T, P, N) system.

4

The Second Law (SL)

This is the longest chapter of the book. It deals with the most important law in thermodynamics. In this chapter I shall also present a new, non-conventional formulation of the Second Law. Before doing that, I would like to begin with a few quotations from Atkins' book (2007).

In the beginning of Chapter 3, on the Second Law, Atkins writes:

> *"...no other scientific law has contributed more to the liberation of the human spirit than the second law of thermodynamics. I hope that you will see in the course*

of this chapter why I take that view, and perhaps go so far as to agree with me."

This is a fancy, pompous statement that is meaningless. First, I have no idea what "liberation of the human spirit" means. But, whatever it may mean, I can assure the reader that it has nothing, absolutely nothing, to do with the Second Law! Reading through Chapter 3 of Atkins' book (T4L), I could see clearly why Atkins holds such a meaningless view. Sadly to say, that this view follows from totally misunderstanding the Second Law of Thermodynamics. I urge the reader to read carefully Chapter 3 of T4L. If he or she agrees with his view please write and explain to me why. Thanks.

On the same page, we find more of the same:

*"The second law is of central importance in the whole of science, and hence in our rational understanding of the universe, because it provides a foundation for understanding why **any** change occurs. Thus, not only is it a basis for understanding why engines run and chemical reactions occur, but it is also a foundation for understanding those most exquisite consequences of chemical reactions, the acts of literary, artistic, and musical creativity that enhance our culture."*

As we shall see in this chapter, the abovementioned statement is not only extremely exaggerated; it is also, and more importantly, untrue. At the moment of writing this book the election of a president of the USA is going on. This will change the future of the USA as well as of the entire world. Does the Second Law explain, or "provide foundation for understanding" these changes?

Contrary to what Atkins claims, the Second Law has nothing to say about "the acts of literary, artistic, and musical creativity."

Regarding the central concept of entropy, Atkins writes:

> *"To fix our ideas in the concrete at an early stage, it will be helpful throughout this account to bear in mind that whereas U is a measure of the **quantity** of energy that a system possesses, S is a measure of the **quality** of that energy; low entropy means high quality; high entropy means low quality."*

This is simply not true! The entropy has nothing to do with "the quality of energy." In fact, there is no such concept in physics that distinguishes between "high quality" and "low quality" of energy. I doubt that such a distinction can be made for any energy. The whole idea of "quality of energy" is a result of misinterpretation of entropy in terms of disorder, and disorder as a "low quality of energy" or "degraded energy". All pure nonsense!

And finally, on page 50 of Atkins' book, we find:

> *"All of our actions, from digestion to artistic creation, are at heart captured by the essence of the operation of a steam engine."*

Nowhere in Atkins' book do you find anything about the relationship between "artistic creation," and "steam engine." As I have explained in my earlier book, Ben-Naim (2016), such statements are not only untrue, but they contribute to deepen the mystery associated with the entropy and the Second Law.

4.1 INTRODUCTION

The Second Law (SL) of thermodynamics is the most interesting, most important, and the most useful law there is in thermodynamics. On the other hand, there is no other law of physics that has been rendered so black and blue as a result of misconstruing, misinterpretation, misunderstanding, and misuse. On top of all these the entropy has been enveloped in a thick cloud of mystery. Some of the reasons for this were discussed in my previous books (Ben-Naim, 2008, 2012, 2016a, 2016b). I will not repeat them here. Instead, I will present in this chapter a brief and concise approach to the SL which I hope will wipe out any misunderstanding, as well as any mystery from the SL.

To be able to do this, I should say from the outset what I believe is essential in order to understand entropy and the Second Law.

1. In order to understand entropy, it is *absolutely necessary* to be familiar with *Shannon's Measure of Information* (SMI).
2. In order to understand SMI, it is *absolutely necessary* to be familiar with the concept of probability.
3. Once you gain understanding of SMI and entropy you will find out that it is *absolutely necessary* to make a clear-cut distinction between these two terms. Failure to do so is the root cause of so much confusion that has prevailed in both thermodynamics and information theory. In addition, once you grasp the meaning of the entropy, you will see (I promise!) that entropy *does not* say anything about the messiness of

my (or your) desk, it *does not* even come close to explain anything about "the acts of literary, artistic, and musical creativity," and, of course, nothing about the fate of the universe!

In this chapter we briefly introduce the concept of probability and probability distribution (PD). Next, we define the SMI on any PD. This is not only important, but a useful and beautiful concept as well, which I believe any educated person should be familiar with. Once we know what SMI is, and understand its meaning, then the concept of entropy will land into our hands like a luscious fruit waiting to be devoured. There will no longer be any mystery associated with this concept. You will also see clearly when and where entropy can be applied, and why numerous scientists, even Nobel Prize recipients, have misapplied this concept.

4.2 THE VARIOUS FORMULATIONS OF THE SL

In this section, we start with a few historical landmarks in the development of the SL. As we have noted in Chapters 2 and 3, the Zeroth Law is associated with the concept of temperature. The First Law introduced the concept of internal energy. Likewise, the SL introduced the concept of entropy — a new concept which did not feature in any branch of physics. In fact, most textbooks of thermodynamics use what we shall refer to as the entropy formulation of the SL. Unfortunately, most authors who use this formulation of the SL fall into the trap created unintentionally by Clausius who coined the term entropy, and concluded that[7]:

"The entropy of the universe always increases."

Another variation of this statement of the SL is: "The entropy tends to increase." The title of Chapter 3 in Atkins' book is: "The Second Law: The increase in entropy."

All these formulations are unfounded and incorrect — more precisely, they are meaningless. First, one must specify the *system* for which the entropy is claimed to increase. Thus, the entropy by itself cannot be said to increase or decrease. Second, the state of the universe is not well defined, therefore the entropy, as a *state function* is not defined for the universe (nor for the "world" as stated by Clausius. (See Note 7)). It is meaningless to say that the *entropy* of the universe increases, or decreases, as it is meaningless to say that the *beauty* of the universe increases or decreases. Similarly, it is as meaningless to say that "entropy tends to increase" as "beauty tends to increase."

We propose in this chapter a new formulation of the SL based on *probability*, rather than on entropy. This is a radically different formulation of the SL and has its origin in Boltzmann's writings. To the best of my knowledge neither Boltzmann, nor anyone else has *formulated* the SL based on probability. Therefore, I will refer to this as the Ben-Naim formulation of the Second Law. As I will demonstrate in the succeeding sections of this chapter, this formulation is far more general than any other formulations of the SL, not only more general, but it is also simpler and easier to comprehend, and does not evoke any mystery as does entropy.

In Section 4.4, we describe the probability formulation of the SL using some simple processes. Once we are done with this formulation, we shall turn to other formulations of the Second Law in terms of entropy, Helmholtz energy, and Gibbs energy.

Because entropy is a relatively new concept in physics, we shall devote Section 4.5 to explaining how the entropy is derived from a more basic and simple-to-comprehend concept of the Shannon Measure of Information (SMI). This approach provides not only a *definition* of entropy, but also an *interpretation* of entropy, and a way of *calculating* values of the entropy.

After presenting the entropy formulation and seeing its limitations, we turn to the equivalent formulations of the SL for a system at constant temperature or the (T, V, N) system. This formulation will be referred to as the Helmholtz energy formulation. Finally, we shall also discuss the Gibbs energy formulation of the SL applicable to (T, P, N) system, i.e. a system at constant temperature and pressure.

As we shall see, the latter two formulations, although formally equivalent to the entropy formulation, are in practice far more useful than the entropy formulation.

4.3 PROBABILITY AND PROBABILITY DISTRIBUTION

Probability theory is a relatively new branch of mathematics. Its roots can be traced back to the 17th century when people pondered on the possibility of "predicting" their chances of success in various games of chance. The penetration of probability into physics was slow and controversial. Nowadays, it is imbued in almost any branch of physics, particularly in statistical mechanics and quantum mechanics.

Notwithstanding its central role in physics, as well as in all other sciences, a definition of probability has been persistently elusive. In this section, we assume that the reader has an intuitive feeling of what probability means. For an elementary discussion of probability, its history, applications and meaning the reader is referred to Ben-Naim (2015).

Next, we shall apply the concept of probability to a system of particles in a box. We start with N indistinguishable particles confined initially in a compartment of volume V. We remove a partition separating the two compartments and observe what happens, (Fig. 4.1).

We shall assume that the particles are simple, i.e. having no internal structure, and all their energy is the kinetic energy associated with the velocity of the particles. We also assume that there are no intermolecular interactions between the particles, and that the particles are indistinguishable.

As we have discussed in Chapter 1, a complete microscopic description of a system of N particles requires at least $6N$ coordinates; three spatial coordinates,

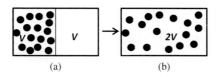

Figure 4.1 Expansion of an ideal gas from V to $2V$.

and three velocity coordinates for each particle. Here, we will assume that in the process depicted in Figure 4.1 there is no change in the *velocity distribution*, or equivalently, no change in the temperature of the (macroscopic) system. Furthermore, we will not be interested in the exact location of each particle but only in which compartment each particle is. Later, in Section 4.1.1, we shall discuss a more general description of the system.

We define the specific configuration of the system when we are given, E, $2V$, N, and in addition we know which *specific* particles are in the right compartment (R), and which specific particles are in the left compartment (L). The *macroscopic* description of the same system is $(E, 2V, N; n)$, where n is the *number* of particles in the compartment L. Thus, in the first description, we are given a *specific configuration* of the system as if the particles were labeled: $1, 2, \ldots, N$. Here, by *specific configuration* we mean only which particles are in R and which are in L. In the *generic* description of the configuration we are given the information only on the *number* of particles in each compartment.

Clearly, if we know only that there are n particles in L, and $N - n$ particles are in R, we have many *specific* configurations that are consistent with the requirement that there are n particles in L.

We denote by $W(n)$ the number of *specific* configurations consistent with n particles in L. Since there are no interactions between the particles we can assume that *all specific* configurations of the system are *equally probable*. Clearly, the total number of *specific* configurations is 2^N, i.e. each particle can be in either one of the two compartments. Using the *classical* definition of the probability, we can calculate the probability of finding n particles in L and $(N - n)$ particles in R. We denote this probability by $P_N(n)$.[8]

It is easy to show that both $W(n)$ and $P_N(n)$ have a maximum as a function of n at the point $n^* = \frac{N}{2}$. (See below). The maximum value of the probability $P_N(n)$ (obtained at $n^* = \frac{N}{2}$), is denoted by $P_N(n^*)$.

Thus, for any given N, there exists an n, such that the number of configurations, $W(n)$, or of the probability, $P_N(n)$ is maximum. Therefore, if we prepare a system with any initial distribution of particles n, and $N - n$ in the two compartments, and let the system evolve, the system will change from a state of lower probability to a higher probability. As N increases, the value of the maximum number of configurations $W(n^*)$ *increases* with N. However, the value of the maximum probability $P_N(n^*)$ *decreases* with N.

To appreciate the significance of this fact, we will examine the "evolution" of systems with small numbers of particles. Later, we shall discuss systems with very large number of particles. We shall see in what sense the spontaneous process of expansion seems to proceed in "one direction only," or is "irreversible."

The case of two particles: $N = 2$

Suppose we have the total of $N = 2$ particles. In this case, we have the following possible configurations and the corresponding probabilities.

$$n = 0 \qquad n = 1 \qquad n = 2,$$
$$P_N(0) = \frac{1}{4}, \qquad P_N(1) = \frac{1}{2}, \qquad P_N(2) = \frac{1}{4}$$

The set of three numbers in the second row is referred to as the PD of this experiment. These numbers mean that if we take snapshots of the configurations of the system, on the average we can expect to find the configuration $n = 1$ (i.e., one particle in each compartment) about half of the time, but each of the configurations $n = 0$ and $n = 2$ only a quarter of the time, Figure 4.2a. If we start with all the particles in the left compartment, we shall find that the system will "expand" from V to $2V$.

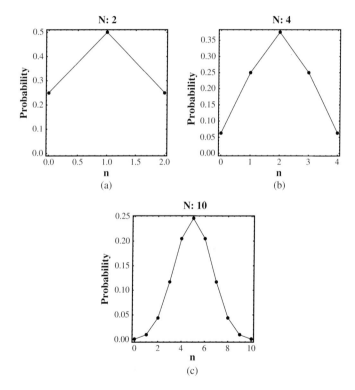

Figure 4.2 Probability of observing n particles in one compartment, and N-n in the other for different numbers N.

However, once in a while the two particles will be found in one compartment. Note that if we run the movie of the evolution of this system, we could not tell whether it is running forward or backwards. The process is completely reversible. There is no sense that the process evolves in one direction. [For simulations of this and similar processes, see Ben-Naim (2008, 2010)].

The case of four particles: N = 4

For the case $N = 4$, we have the PD as shown below in Fig. 4.2b.

$n = 0$	$n = 1$	$n = 2$	$n = 3$	$n = 4$
$P_N(0) = \dfrac{1}{16}$	$P_N(1) = \dfrac{1}{4}$	$P_N(2) = \dfrac{3}{8}$	$P_N(3) = \dfrac{1}{4}$	$P_N(4) = \dfrac{1}{16}$

The maximum probability is $P_N(2) = \frac{3}{8} = 0.375$ which is *smaller* than $\frac{1}{2}$. In this case, the system will spend only $\frac{3}{8}$ of the time in the maximum state $n^* = 2$. Again, if we start with all particles in one compartment, the system will "expand" from V to $2V$, but once in a while we shall see all the particles in one compartment. Again, running the movie of this system will not tell us whether it runs forward or backwards. There should be no sense of "evolution in one direction" only.

The case of ten particles: N = 10

For $N = 10$, the distribution is shown in Figure 4.2c. We calculate the maximum at $n^* = 5$ which is $P_{10}(n^* = 5) = 0.246$.

In all of the three examples we examined above, the system *expands* from V to $2V$. However, there is nothing in this process which can be termed "irreversible." In each case, the initial state will be visited once in a while, if we only wait for a sufficiently long period of time. [For some simulations of the evolution of such systems, see Ben-Naim (2008, 2010).]

Very large number of particles

Let us proceed with larger N. Figure 4.3 shows the probabilities $P_N(n)$ for larger number of particles. It is seen that the maximum value of $P_N(n)$ *decreases* as N *increases*.

Note also that the probability distribution $P_N(n)$ becomes closer to the Normal distribution when N is large. Now, suppose that we start with all particles in one compartment (either $n = 0$ or $n = N$), and run the movie showing how the configuration changes with time, we shall see that the system will *expand* from V to $2V$. If N is 1000 or more, we shall never observe any "visit" to the original configuration. We might conclude that the process is *irreversible*, i.e. it has a one-direction of evolution. However, this sense of "directionality" is only an illusion. If we run the movie for billions of billions of years, we will observe "visits" to the original configuration, and we will have no sense of an irreversible process occurring in the system.

It can be shown that the maximum probability *decreases* as $N^{-1/2}$. In practice, we know that when the system reaches the state of equilibrium, it stays there *forever*. The reason is that the *macroscopic state of equilibrium* is not the same as the state for which $n^* = \frac{N}{2}$, but it is this state along with a small neighborhood of n^*, say $n^* - \varepsilon N \leq n \leq n^* + \varepsilon N$, where ε is a small number. For $N = 100$ and $\varepsilon = 0.01$, the probability of finding n in the neighborhood of n^* is

82　*The Four Laws That Do Not Drive The Universe*

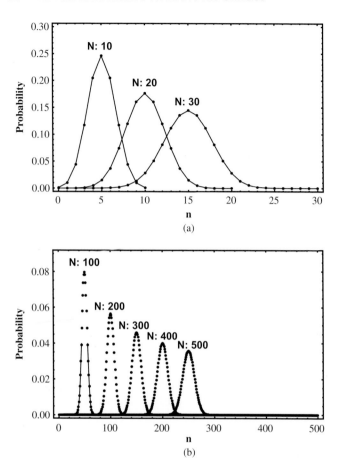

Figure 4.3 Probability of observing n particles in one compartment, and N-n in the other.

about 0.235. For $N = 10^{10}$ particles, we can allow deviations of 0.001% of N and the probability of staying in this neighborhood is nearly one. [For more details, see Ben-Naim (2008, 2012).]

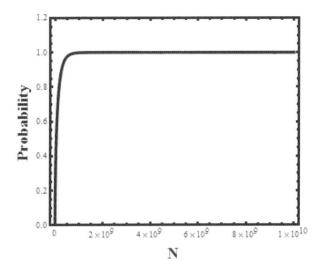

Figure 4.4 The probability of finding n particle in the neighborhood of $n^* = N/2$, in one compartment as a function of N, with $\varepsilon = 0.0001$.

In Fig. 4.4, we show the probability of finding n between $n^* - \varepsilon N \leq n \leq n^* + \varepsilon N$ as a function of N, with $\varepsilon = 0.0001$. Plotting the probability $P_N(n^* - \varepsilon N \leq n \leq n^* + \varepsilon N)$ as a function of N shows that this probability tends to *one* as N increases. When N is on the order of 10^{23}, we can allow deviations of $\pm 0.00001\%$ of N, or even smaller, yet the probability of finding n at or near n^* will be almost one. It is for this reason that when the system reaches n^* or near n^*, it will stay in the vicinity of n^* for most of the time. For N on the order of 10^{23}, "most of the time" means practically, *always*.

The abovementioned specific example provides an explanation for the fact that the system will "always" evolve in "one direction," and "always" stays at the equilibrium state once that state is reached. The tendency towards a state of larger probability is equivalent to the statement that events that are supposed to occur more frequently will occur more frequently. This is plain common sense. The fact that we do not observe deviations from the monotonic climbing of n towards n^*, or for n staying close to n^*, is a result of our inability to detect small changes in n-(or equivalently small changes in the SMI, see below). Note that in this section we did not say anything about the entropy changes. Before turning to calculate the entropy changes we repeat the main conclusion of this section. For each N the probability of finding a distribution of particles: $(n, N - n)$ in the two compartments L and R has a maximum at $n^* = \frac{N}{2}$. For a very large number of particles the probability of obtaining the *exact* value of $n^* = \frac{N}{2}$ is not very large. However, the probability of finding the system at a small vicinity of $n^* = \frac{N}{2}$ is almost one! See Fig. 4.4.

When we say that the system has reached an equilibrium state we mean that we do not *see* any changes that occur in the system. In this example, we mean changes in the *density* of the particles in the entire system.

At equilibrium the macroscopic density that can be measured at each point in the system is constant. In the particular system we discussed above the measurable density of the particles in the two compartments is $\rho^* \cong N/2V$. Note that fluctuations always occur. Small fluctuations occur very frequently, but they are so small that we cannot measure them. On the other hand, fluctuations that could have been measured are extremely infrequent, and practically we can say that they never occur. This conclusion is valid for very large N.

4.4 THE PROBABILITY FORMULATION OF THE SECOND LAW

Let us state the SL for this particular system (generalizations will follow in the subsequent sections). We start with a system of N particles in one compartment, where N is of the order of one Avogadro number, about 10^{23} particles. We remove the partition and follow the evolution of the system. At any point in time we define the *distribution* of *particles* by the pair of numbers $(n, N - n)$. Of course, we do not count the *exact* number of particles in each compartment n, but we can measure the density of particles in each compartment, $\rho_L = n_L/V$ and $\rho_R = n_R/V$, where n_L and n_R are the numbers of particles in the left (L) and right (R) compartments, respectively ($n_L + n_R = N$). From the measurement of ρ_L and ρ_R we can also calculate the pair of *mole fractions* $x_L = n_L/(n_L + n_R) = \rho_L/(\rho_L + \rho_R)$ and $x_R = n_R/(n_L + n_R) = \rho_R/(\rho_L + \rho_R)$, with $x_L + x_R = 1$. The pair of numbers (x_L, x_R) is referred to as the *configuration* of the system. Note that the pair (x_L, x_R) is also a probability distribution. Can you explain why the pair of numbers (x_L, x_R) is a PD?[9]

After the removal of the partition between the two compartments, we can ask what the probability of finding the system with a particular configuration (x_L, x_R) is. We denote this probability by $\Pr(x_L, x_R)$. Since both x_L and Pr are probabilities, we shall refer to the latter as *super* probability; $\Pr(x_L, x_R)$ is the probability of finding the PD (x_L, x_R). We can now state the SL for this particular system as follows:

Upon the removal of the partition between the two compartments, the PD, or the *configuration* will evolve from the initial one $(x_L, x_R) = (1, 0)$, (i.e. all particles in the left compartment) to a new equilibrium state characterized by a uniform locational distribution. This means that the densities ρ_L and ρ_R are equal (except for negligible deviations), or equivalently the mole fractions x_L and x_R are equal to ½. We shall *never* observe any significant deviation from this new equilibrium state, not in our lifetime, and not in the universe's lifetime which is estimated to be about 15 billion years.

Note that *before* we removed the partition the probability of finding the configuration (1, 0) is *one*. This is an equilibrium state, and all the particles are, by definition of the initial state, in the L compartment.

We refer to the *super probability* of finding the configuration (x_L, x_R), denoted by $\Pr(x_L, x_R)$, as the probability of the configuration attained *after* the removal of the partition when x_L can, in principle, attain any value between zero and one. Therefore, the super probability of obtaining the configuration (1, 0) is negligibly small. On the other hand, the super probability of obtaining the configuration in the neighborhood of $\left(\frac{1}{2}, \frac{1}{2}\right)$ is, for all practical purposes nearly one. This means that *after* the removal of the partition, and reaching an equilibrium state, the ratio of the super probabilities of the initial configuration (1, 0) and the final configuration, i.e. in the neighborhood of $\left(\frac{1}{2}, \frac{1}{2}\right)$, is almost infinity (of the order of 2^N) with $N \approx 10^{23}$, this is an unimaginably large number). Thus, we can say that for 10^{23}

$$\frac{\Pr(\textit{final configuration})}{\Pr(\textit{initial configuration})} \approx \textit{infinity}$$

This is the probability formulation of the Second Law for this particular experiment. This law states that starting with an equilibrium state where all particles are in L, and removing the constraint (the partition), the system will evolve to a new equilibrium configuration which has a probability overwhelmingly larger than the initial configuration.

Note carefully that if N is small, then the evolution of the configuration will not be monotonic, and the ratio of the super probabilities in the equation above is not near infinity. For some simulations the reader is referred to Ben-Naim (2008, 2010). For very large N, the evolution of the configuration is also not strictly monotonic, and the ratio of the super probabilities is not strictly, infinity. However, in *practice*, whenever N is large we shall *never* observe any deviations from monotonic change of the configuration from the initial value $(1, 0)$ to the final configuration $\left(\frac{1}{2}, \frac{1}{2}\right)$. Once the final equilibrium state is reached [i.e. that the configuration is within experimental error $\left(\frac{1}{2}, \frac{1}{2}\right)$], it will stay there *forever*, or equivalently it will be found with probability one.

The distinction between the strictly *mathematical* monotonic change and the *practical* change is important. The process is mathematically *always* reversible, i.e. the initial state will be visited. However, in practice the process is irreversible; we shall *never* see the reversal to the initial state.

Let us repeat the probability formulation of the Second Law for this particular example.

We start with an initial *constrained equilibrium state*. We remove the constraint, and the system's configuration will evolve with probability (nearly) one, to a new equilibrium state, and we shall *never* observe reversal to the initial state. "Never" here, means never in our lifetime, nor in the lifetime of the universe.

This formulation is valid for large N. It is also valid for any *initial constrained equilibrium state*. As we will see in the next few sections the entropy formulation, the Helmholtz energy formulation, and the Gibbs energy formulation pertain to specific thermodynamic systems; isolated (E, V, N), isothermal (E, V, N), and isothermal isobaric (T, P, N), respectively. In this sense, the probability formulation is very general as it applies to any thermodynamic system.

4.4.1 Generalizations of the Probability Formulation of the Second Law

We next generalize our discussion of the previous section to more general processes. We do this in three steps:

(a) *Removal of constraints on the locations of the particles*

In the previous section we discussed the expansion from V to $2V$. Initially, all the particles were constrained to a volume V and, after removal of the partition the particles expanded to occupy the entire volume of $2V$.

Figure 4.5 shows three boxes (in 2D) each of total volume V. Initially, the particles in each box are distributed in c compartments, such that the particle distribution is

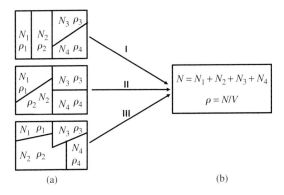

Figure 4.5 (a) Three different constrained equilibrium systems, all having the same total E, V and N. *The number densities are different in each compartment.* (b) The unconstrained equilibrium state obtained after the removal of the constraints. The final density is uniform throughout the system $\rho = N/V$.

$(N_1, N_2, ..., N_c)$, i.e. N_i particles are in compartment i. We assume again that the particles do not interact with each other (ideal gas), and that the partitions between the compartments are heat conducting so that the temperature throughout the system is constant. (We shall remove this restriction in the next sub-section).

For each *particle distribution* $(N_1, N_2, ..., N_c)$ correspond a *probability distribution* $(x_1, x_2, ..., x_c)$ where x_i is the mole fraction of particles in compartment i (we refer to $(x_1, x_2, ..., x_c)$ as the initial *configuration* of the system.

The Probability formulation of the SL for this system is as follows:

For any *initially constrained equilibrium state* as shown on the left hand side of Fig. 4.5, when we remove the constraints, the system's configuration $(x_1, x_2, ..., x_c)$ will evolve with probability nearly one, to a new equilibrium state such that the density throughout the entire system is constant.[10]

Initially, in each compartment the density was $\rho_i = N_i/V_i$. After the removal of all the partitions the density in each (virtual) compartment will be constant $\rho_i = N/V$ (where $N = \Sigma N_i$ and $V = \Sigma V_i$).

(b) Removal of constraints on the temperatures

We start with an initial system similar to that on the left hand side of Fig. 4.5, but now we prepare the initial *particle*-distribution such that the density in each compartment is the same $\rho_i = \rho = N/V$. However, the temperature in each compartment is different, say $(T_1, T_2, ..., T_c)$. The partitions between the compartments are insulators; no heat transfer is allowed. (Fig. 4.6).

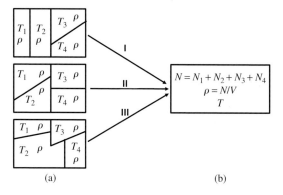

Figure 4.6 (a) Three different constrained equilibrium systems, All having the same total E, V and N. *The number densities are the same in each compartment, but the temperatures are different* (b) The unconstrained equilibrium state obtained after the removal of the constraints. The final temperature is uniform throughout the system.

The probability formulation of the SL for this system is:

When we remove the constraints, the system will evolve with probability of nearly one to a new equilibrium state for which the temperature will be uniform throughout the entire system.

We note that we use here the temperature to characterize each compartment. Underlying the temperature characterization is a more fundamental characterization of the system in terms of the velocity distribution of the particles. We shall describe this distribution in Section 4.5.2.

(c) Removal of constraints on both the locations and the temperatures

The final generalization of the probability formulation of the SL is shown in Fig. 4.7. Here, we initially have an

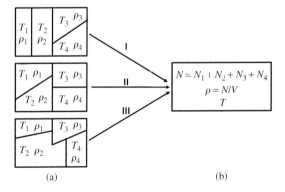

Figure 4.7 (a) Three different constrained equilibrium systems, all having the same total E, V and N. *The number densities and the temperatures are different in each compartment.* (b) The unconstrained equilibrium state obtained after the removal of the constraints. The final density and temperature are uniform throughout the system.

arbitrary distribution of particles, as well as arbitrary distribution of temperatures in the compartments. We still assume that N is large and the particles do not interact with each other (ideal gas). The probability formulation of the SL for this case is:

We start with any particle distribution $(N_1, N_2,..., N_c)$, and any temperature distribution $(T_1, T_2,..., T_c)$ in a constrained equilibrium state (i.e. no change in the density ρ_i of each compartment and no change in the temperature T_i). We remove the constraints and the system will evolve with probability nearly one to a new equilibrium state characterized by a uniform density $\rho_i = \rho = N/V$ and uniform temperature $T_i = T$.

Before we proceed to the entropy formulation of the SL we note that in the three cases discussed above we have characterized the system in both the initial and the final states by macroscopic (measurable) variables; the densities ρ_i and the temperatures T_i of each compartment. In Section 4.5.1, we shall discuss the characterization of the configurations in terms of distribution of *locations* and *velocities* of all the particles. The latter description is more general than the former. However, while we can prepare an initial constrained equilibrium state with any arbitrary densities $(\rho_1,..., \rho_c)$ and temperatures $(T_1,..., T_c)$ we cannot prepare an equilibrium state with arbitrary distribution of locations and velocities.

Note carefully that we formulated the SL in terms of the probability *without* mentioning entropy. This is in sharp contrast with most formulations of the Second Law.

Pause and ponder.

So far we formulated the Second Law in terms of probabilities. We shall soon see that this formulation is equivalent to the entropy formulation of the Second Law. At this stage I suggest that you return to the beginning of this chapter, read the few quotations from Atkins' book and answer the following questions:

Can you explain how our thoughts and feelings or creation of art follow from the Second Law?

Is the sunrise that you observe in the morning a result of the Second Law?

Can the expansion of the entire universe be understood in terms of the Second Law?

If you have an answer to any of these questions please write to me and I will add your answer to the second edition of this book.

4.5 THE ENTROPY FORMULATION OF THE SECOND LAW

In this section, we present the traditional formulation of the SL in terms of the entropy. It should be emphasized from the outset that this particular formulation applies only to *isolated* systems, and in this sense it is less general when compared to the probability formulation of the Second Law.

Most authors of books on entropy and the SL fail to see the limitation of the applicability of the entropy in entropy formulation, ending up with all sorts of fancy, pompous and meaningless statements regarding the "tendency of entropy to increase," "the entropy of the universe," "the entropy of living systems," and many more.

Entropy may be defined, as Clausius did, by means of experimental quantities. However, this definition though useful does not offer any interpretation of the entropy. Therefore, in the next subsection we further develop the concept of entropy based on Shannon's Measure of Information (SMI).

In my opinion, there is no way of understanding the *meaning* of entropy without understanding the meaning of the SMI. In other words, familiarity with the concept of SMI is indispensable for the understanding entropy. In what follows, we shall be very brief. For a more detailed discussion of the SMI and its relevance to entropy the reader is referred to literature listed in the bibliography.

4.5.1 Application of the SMI to a Thermodynamic System

Given any PD distribution $(p_1, p_2, \ldots, p_n)$ where p_i is the probability of the outcome i for an experiment, we define the SMI on this distribution by

$$\text{SMI} = H(p_1, p_2, \ldots, p_n) = -\sum p_i \log p_i$$

where log is the logarithm with base 2, and the summation is over all possible outcomes of the experiment.

Note that here we defined the SMI for an experiment with a finite number of outcomes. One can also define the SMI on a continuous PD. We shall not need this in the present elementary and qualitative discussion. As can be seen from its definition, the SMI has the structure of an average. Normally, an average quantity is obtained by taking the sum of the products of $p_i A_i$ where A_i is some number associated with the outcome *i*. In the case of SMI, we have a very special average. The quantities A_i are $\log p_i$, therefore, the SMI is a purely probabilistic quantity; it is a function of the *whole probability distribution*.

4.5.2 Definition of the SMI on the Probability Distribution of Locations and Velocities of the Particles

Now that we have the notion of SMI we apply it to a system of particles (initially an ideal gas).

Suppose that we could take snapshots of the microscopic configurations of the systems at many points in time. Each snapshot records all the *locations* and all the *velocities* of the particles. Theoretically, we have infinite possibilities for the locations and the velocities of each particle. However, in *practice* there is a limit on the accuracy of determining the location and velocity of each particle.[11]

Thus, instead of an infinite number of possible configurations we have only a finite number of configurations. Each configuration of the entire system is now viewed as a possible outcome of the experiment (the experiment is taking the snapshots to determine the configurations).

We next define the probability distribution of these outcomes, $p_1, p_2, ..., p_n$, where p_i is the probability of finding the configuration i [which is the specification of all the locations and all the velocities of the particles within a small "cell" of "volume" $(dxdydzdv_x dv_y dv_z)$].

Note that each characterization of a constrained equilibrium state, discussed in the previous section, corresponds to many locational configurations of the particles. For instance, if we have four compartments with particle distribution (N_1, N_2, N_3, N_4), then such a macroscopic characterization corresponds to many molecular characterizations of the locations of the particles.

Clearly, the molecular characterization is far more detailed than the macroscopic or thermodynamic characterization of the configuration.

For each probability distribution we can define the SMI. At the same time we can also ask: What is the super probability of finding a specific probability distribution, which we denote by $\Pr(p_1, p_2, ..., p_n)$. It turns out that the SMI, defined on the probability distribution $(p_1, p_2, ..., p_n)$, and the super probability defined on the same distribution are related to each other. Also, one can prove that that there exists *one distribution* which maximizes the SMI, as well as the super probability Pr. For details, see Ben-Naim (2008, 2012, 2016).

4.5.3 Definition of Entropy

Historically, the term *entropy* was coined by Clausius. In fact, Clausius defined only changes in entropy (dS) for a particular process of transferring a small amount of heat δQ to a system at constant temperature $dS = \delta Q / T$. Later,

Boltzmann defined the so-called "absolute entropy" of an isolated system by $S = k_B \log W$, where k_B is the Boltzmann constant, and W is the number of accessible microstates of a thermodynamic system. Recently, a new definition of entropy based on SMI was suggested by Ben-Naim [Ben-Naim (2008, 2015a, 2016a, 2016c)]. The latter definition has several advantages that were described in great detail recently. Here, we note that all the three definitions provide the same results for processes for which the entropy changes may be calculated.

In the previous sections, we discussed the concept of SMI. We now apply this concept to the same PD $p_1, \ldots, p_n$ of the configurations of a thermodynamic system. As we noted in the previous section, one can prove that there exists a PD denoted by $p_1^*, \ldots, p_n^*$, and to which we refer to as the *equilibrium* PD which maximizes both the SMI and the super probability Pr.

We now define the entropy of an ideal gas by the *maximum* value of the SMI. It is maximum with respect to all possible probability distributions of locations and velocities.

It is important to distinguish between the concept of SMI which may be defined for *any* probability distribution, and the concept of entropy which is defined for a particular set of distributions. The steps leading from SMI to entropy are as follows:

First, we define the configuration of a system of N particles in terms of the locations and velocities (or momenta) of all the particles.

Second, we define the probability distribution for all possible configurations of a specific thermodynamic system.

Finally, we choose a particular distribution referred to as the *equilibrium distribution*, which turns out to *maximize* both the SMI and the super probability. The value of this maximum SMI, after multiplying by a constant, is the entropy of an ideal gas.

One can extend this definition of entropy to a system of interacting particles. We shall not be concerned with this extension. (See Ben-Naim, 2008).

As a particular case of SMI, the entropy carries the same *interpretation* and *meaning* as that of SMI. This meaning cannot be obtained from either Clausius' or Boltzmann's definitions. This is one of the advantages of the present definition of entropy.

Furthermore, the relationship between the SMI and the super probability discussed above is also valid for the entropy; the maximum value of the SMI which is proportional to the entropy is related to the maximal value of the super probability. This relationship also provides the entropy formulation of the SL.

It should be said that in most popular science books entropy is interpreted as a measure of disorder. Here is a typical example.

In Atkins' book we find on page 61:

> *"For our initial encounter with the concept, we shall identify entropy with disorder…"*
>
> *"With disorder in mind, we shall explore the implications of Clausius's expression and verify that it is plausible in capturing the entropy as a measure of the disorder in a system."*

As I have shown in my previous book, Ben-Naim (2016), entropy *cannot* be identified (neither in an initial nor in a final encounter) with disorder! It is true that in some specific processes, like the mixing of two ideal gases, increase in entropy is correlated with what seems to us an increase in disorder. Unfortunately, this correlation is not always true, and entropy can never be identified with disorder.

Not only is entropy not a measure of disorder, the Second Law does *not* imply a tendency towards more disorder!

In his conclusion of the Second Law, Atkins writes:

"Wherever structure is to be conjured from disorder, it must be driven by the generation of greater disorder elsewhere, so that there is a net increase in disorder of the universe, with disorder understood in the sophisticated manner that we have sketched. That is clearly true for an actual heat engine, as we have seen. However, it is in fact universally true."

Nothing in these sentences is true. What Atkins claims to be "universally true," is in fact, universally untrue, universally meaningless and universally misleading!

4.5.4 The Entropy Formulation of the Second Law

The entropy formulation of the SL applies only to isolated systems. We shall formulate it for a one component system having N particles. If there are k components, then N is reinterpreted as a vector comprising the numbers $(N_1, N_2, ..., N_k)$ where N_i is the number of particles of species i.

For any isolated system (E, V, N), at equilibrium, the entropy is maximum over all possible constrained equilibrium states of the same system.

Note that this formulation uses only macroscopic quantities. Also, it applies only to equilibrium states.

In Section 4.4, we discussed a few constrained equilibrium states of the same system (E, V, N). The entropy formulation means that if we remove any of the constraints in any of the initial systems, the entropy will either increase or remain the same.

Therefore, an equivalent formulation of the Second Law is:

Removing any constraint from a constrained equilibrium state of an isolated system will result in an increase of the entropy.

Note carefully that we have *defined* entropy for equilibrium systems. The maximum entropy is also a maximum with respect to all *constrained equilibrium states*. This is very different from the maximum of the SMI defined in Section 4.5.3.

Recall that the SMI may be defined for any distribution (coin, die, particles in a box, etc.). Also, for a (E, V, N) system of particles, we can define the SMI for *any N* and *any* distribution. We looked for the particular distribution which maximizes the SMI. The value of the maximum SMI is proportional to the entropy of the system.

Look again at Fig. 4.5, 4.6, and 4.7. Any of the constrained equilibrium states may be viewed also as a distribution of locations and momenta of all the particles.

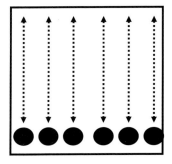

Figure 4.8 A system with particles moving at constant velocity upwards and downwards.

The converse of this sentence is not true. Not every distribution of locations and momenta is a constrained equilibrium state. An example: Take a system of N particles in a box of length L all moving back and forth at constant speed v, (Fig. 4.8). For this system the probability distribution of locations and momenta is well-defined. Therefore, also the SMI is well-defined. If the collisions with the walls are perfectly elastic, the state of this system will not change with time. However, this system *is not* a constrained equilibrium state in a thermodynamic sense. Hence, the entropy is not defined for such a system.

Thus, if we have any division of the system (E, V, N) in which each compartment is well-characterized by a distribution of locations and momenta we can define the corresponding SMI for each compartment.

The entropy of each compartment is related to the maximum SMI of that compartment at equilibrium. The entropy of the entire system (E, V, N) is maximum over all the entropies of the same system being in constrained equilibrium states.

Because of its fundamental importance we repeat the steps leading from SMI to entropy. The SMI is defined for any system of particles, and for any distribution of locations and momenta. There is *one distribution* that maximizes the SMI. The *value* of the maximum SMI up to a multiplicative constant ($k_B \ln 2$) is the entropy of the system. The entropy of the system pertains to the equilibrium state and as such it is *not* a function of time and it does not change with time. Note that the entropy is obtained for the *distribution* which maximizes the SMI. The connection with the equilibrium distribution follows from the following considerations. The distribution which maximizes the SMI is the same distribution that maximizes the super probability (Pr). We know that any system tends to equilibrium. Therefore, we identify the distribution which maximizes Pr as the *equilibrium* distribution. The principle of maximum entropy of an isolated system states that the entropy of the system is maximum over all possible *constrained equilibrium states* of the same system. Each of these constrained equilibrium states has entropy, which is also not a function of time.

The reader might get an impression that "maximum over all constrained equilibrium states" is a more restricted definition than that of "maximum over all possible distributions." Indeed, each "constrained equilibrium state" defines a "molecular distribution of locations and velocities." The inverse of the last sentence is in general not true. However, one should remember that the entropy is defined for equilibrium states, and "maximum over all constrained equilibrium states" is rich enough for all practical purposes. Indeed, the SMI is definable for any molecular distribution, of small or large system, and of equilibrium or non-equilibrium states.

Failing to distinguish between the SMI and entropy has caused a great confusion. It was started by Boltzmann who defined a quantity H, and claimed that it decreases with time and reaches a minimum at equilibrium or at $t \to \infty$. It is unfortunate that the function $-H(t)$ was identified with entropy, and therefore people concluded that the entropy reaches a maximum at $t \to \infty$, i.e. as a function of time.

The truth is that the function $-H(t)$ is SMI defined on the distribution of locations and momenta. Indeed, this function changes with time and reaches a maximum at equilibrium ($t \to \infty$). It is only at the limit $t \to \infty$ that the function $-H(t)$ attains its maximum value, that we can identify its maximum value (of this SMI) with the entropy of the system.

In Fig. 4.9, we schematically show the relationship between the distribution (of locations and momenta), the definition of the SMI, and Pr on each distribution, and the maximal SMI corresponding to the equilibrium distribution. The same equilibrium distribution also maximizes the super-probability. The maximum value of the SMI is related to the entropy of the system. The approach towards a maximum Pr explains the Second Law. The connection between the SMI and Pr is by a monotonic function, hence the maximum SMI corresponds to the maximal Pr.

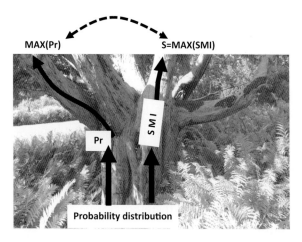

Figure 4.9 The entropy formulation of the Second Law, and its relationship with the probability formulation.

Finally, we can ask what drives the process when we remove a constraint. The answer is that the kinetic energy of the particles is the ultimate *cause* of any change that occurs in the system. Without the kinetic energy of the particles no process will occur, and the system will not evolve towards equilibrium. The *direction* of the change from one equilibrium state to another is determined by the probability.

In this view, it is clear that the entropy change is not the *cause* of the change but rather a *result* of the change.

In most textbooks, you might find formulations of the Second Law as:

The entropy has a tendency to increase.
The entropy of the universe always increases. When it reaches its maximum the universe will be rendered thermally dead.

The entropy of an isolated system increases until it reaches a maximum.

All of the above sound similar, but in fact, they are all wrong: The entropy does not have a tendency to increase! The entropy of the universe is not defined! The entropy of an isolated system is defined for an equilibrium state. As such, it does not "increase until it reaches a maximum!"

Most people who accept either of the above formulations would (unjustifiably) extend them, assigning to entropy the power to *drive* anything, from the expansion of a gas, to breaking and splattering of an egg, to formation of thoughts, to the creation of the arts — in short, to drive the universe! This is the very essence of the title of Atkins' book.

The truth is, entropy does not drive any process and certainly is not even defined for a living system, and *a fortiori* so for the entire universe.

Believe it or not — aside from the misleading title of his book, Atkins writes on page 65:

"A feasible process is spontaneous only if the total entropy of the universe increases."

On page 82, he writes:

"To identify spontaneous processes we must note the crucially important aspect of the Second Law that it refers to the entropy of the universe… Whenever we see the apparently spontaneous reduction of entropy, as when a structure emerges, a crystal forms, a plant grows, or a thought emerges, there is always a greater

increase in entropy elsewhere that accompanies the reduction in entropy of the system."

These may be beautifully crafted words, perhaps even impressive, but at best, hollow.

Consider this "thought": I just thought that as a result of my thoughts my entropy decreased. I wonder whether Atkins knows where in the universe the entropy has increased as a result of my thoughts, *and* as a result of my wondering about the entropy of the universe.

I hope the reader can see the double fallacy of Atkins' statements. First, he assumes that "plant grows" and "thought emerges" etc. are associated with an apparent reduction in entropy — which is untrue. Second, he claims that the entropy of the universe must increase, to compensate for this reduction of entropy — which is equally untrue.

4.6 THE HELMHOLTZ ENERGY FORMULATION OF THE SECOND LAW

In the previous section, we discussed the entropy formulation of the Second Law, and emphasized that it applies to a well-defined isolated system denoted by (E, V, N). Of course, a strictly isolated system does not exist, and even if it did exist, we could not have carried out any measurement on it. Yet, an isolated system is a convenient concept for the development of thermodynamics and statistical mechanics.

A more convenient system is a closed system having a fixed volume V, and constant temperature T.

Such a system can exchange heat with its surroundings. Normally, we keep the temperature of the *system* fixed by placing it in a heated *bath*. The heat bath is supposed to be very large compared with the system, such that when there is a small exchange of heat between the system and the bath, the bath's temperature is not affected.

Before we formulate the SL for such a system we defined the Helmholtz energy by:

$$A = E - TS.$$

Here we have on the right-hand-side of the equation the energy E, the temperature T and the entropy S. Since S is *only* defined for equilibrium systems, so is the Helmholtz energy.[12]

This definition of the Helmholtz energy is valid for any system at equilibrium. In this definition we did not specify the independent variables with which we characterize the system. We have the liberty to choose any set of independent variables we choose, say (E, V, N), (T, V, N) or (T, P, N). However, if we want to formulate the SL in terms of the Helmholtz energy we have no other choice but to choose the independent variables (T, V, N). For a (T, V, N) system the Helmholtz energy formulation of the Second Law is:

For any (T, V, N) system at equilibrium the Helmholtz energy has a minimum over all possible constrained equilibrium states of the same system.

Recall that the entropy formulation of the SL was valid only for *isolated* systems, i.e. for (E, V, N) systems. The Helmholtz energy formulation is valid only systems that are isothermal (constant T), and isochoric (constant V), as well as closed (constant N).

An equivalent statement of the Helmholtz energy formulation is:

Removing any constraint from a constrained equilibrium state in a (T, V, N) system will result in a decrease in Helmholtz energy.

Note carefully that it is only for a process at constant (T, V, N) that this formulation is valid. It is not true that the Helmholtz energy decreases for any process occurring in any thermodynamic system.

The Helmholtz energy formulation is equivalent to the entropy formulation in the following sense.

Mathematically, we start with the *entropy function* $S(E, V, N)$ and make a transformation of variables from (E, V, N) to (T, V, N). One can prove that the minimum of the Helmholtz energy *function* $A(T, V, N) = E(T, V, N) - TS(T, V, N)$ follows from the maximum entropy. This transformation is known as the Legendre transformation.[13] An excellent exposition of this transformation is provided in Callen's book (1985).

The reverse of the last statement is also true; i.e. from the minimum principle of the Helmholtz energy function one can get the maximum entropy principle provided that we transform back from the (T, V, N) to the (E, V, N) variables.

Thus, you can look again at the processes in Fig. 4.5 to 4.7, and replace all the E, V, N systems by (T, V, N) systems, and the Helmholtz energy principle will apply.

Because the Legendre transformation is mathematical we do not bring it here. Instead, we present a qualitative

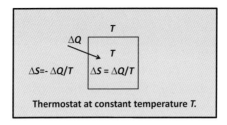

Figure 4.10 A system in contact with an isolated heat reservoir (a thermostat).

argument based on the experiment shown in Fig. 4.10. Here, we have a (T, V, N) system immersed in a very large thermal bath. We allow exchange of heat between the system and the bath, but no exchange of volume (i.e. constant V), and no exchange of matter (constant N). The combined system plus the bath is isolated, and will be referred to as *total*.

Now suppose we remove a constraint in the equilibrated system (this could be a removal of the partition or adding a catalyst which enables chemical reactions).

Suppose that as a result of the process in the system, the entropy of the system has changed, denote this change by $\Delta S(sys)$. This change can be either positive or negative. Also, suppose that the energy of the system changed and denote this change by $\Delta E(sys)$. Again this change can be either positive or negative.

Since the process was carried out at constant T, and since the total energy of the system plus bath is constant, we must have the equality for the combined system plus bath, i.e., for the *total*

$$\Delta E(total) = \Delta E(sys) + \Delta E(bath) = 0$$

The system and the bath can exchange only heat (no work). Therefore, the change $\Delta E(sys)$ must be a result of the flow of heat $\Delta Q(bath \rightarrow system)$. If $\Delta E(sys)$ is positive, then ΔQ is positive (heat flows from the bath into the system). If $\Delta E(sys)$ is negative, then ΔQ is negative.

Now, we use Clausius' principle. Since the bath exchanges heat with the system, the entropy change of the bath is $\Delta S(bath) = -\Delta Q/T$. If ΔQ is positive, i.e. heat flows into the system, then $\Delta S(bath)$ will be *negative*. We still cannot say anything about the sign of the change in the entropy of the system. The reason is that the system's entropy has changed both because of the heat flow and the processes that occurred following the removal of some constraints.

The system, plus bath which we refer to as *total*, is isolated. Therefore, the entropy change for the *total* must be positive, i.e.

$$\Delta S(total) = \Delta S(sys) + \Delta S(bath)$$
$$= \Delta S(sys) - \Delta Q/T = \Delta S(sys) - \Delta E(sys)/T \geq 0$$

Rearranging this equation we obtain

$$-T\Delta S(total) = \Delta E(sys) - T\Delta S(sys) = \Delta A(sys)$$

This is a remarkable equation. The entropy change of the *total* (system plus bath) times the temperature T is equal to minus the Helmholtz energy change of the *system*. Although we do not know the values of the changes $\Delta E(sys)$ and $\Delta S(sys)$, each can be either positive or negative. We can say that $\Delta S(total)$ is positive (because of the entropy formulation of the SL for the isolated *total*). Therefore, we can conclude that the sign of $\Delta A(sys)$ must be negative. Thus we have shown that the Helmholtz energy formulation, for the (T, V, N) system, follows from the entropy formulation for the (E, V, N) system.

It is absolutely important to remember the assumptions that we made in deriving this conclusion: First, that the process in the system occurred at constant T, V, N. The variables V and N of the system were kept constant by means of rigid and impermeable walls. The temperature of the system was kept constant by requiring that the bath be so large that even when it exchanges heat (ΔQ) with the system its temperature is not affected. Second, and most importantly the system + bath is isolated. Therefore, we could apply the entropy formulation to the *total*. From this entropy formulation we derived the Helmholtz energy formulation for the (T, V, N) system. Third, we assume that the bath can only exchange heat with the system. No other processes is presumed to take place in the bath.

Exercise: Before we continue, pause and think. Can you repeat the same argument as above, but instead of the bath at constant T, the system is in the thermal contact with the *universe*, (Fig. 4.11). We do the same process as before, i.e. we remove some constraints from the system. If you are not sure, I suggest that you re-read the entire section and convince yourself that the conclusion we reached is true *provided* that the temperature of the bath is constant and that the system plus the bath is an isolated system. After this, I suggest you read pages 83–85 of Atkins' book (2007). You will see that Atkins did exactly this exercise and reached the absurd conclusion that:

112 *The Four Laws That Do Not Drive The Universe*

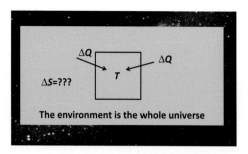

Figure 4.11 A system in contact with the entire universe.

"A change in A is just a disguised form of the change in the total entropy of the universe when the temperature and the volume of the system are constant."

The change in A is *not* a disguised form of the change in the total entropy, and the entropy of the universe is *not* defined! From this statement, one can conclude that Atkins' book is a disguised form of a popular science book!

Atkin's fatal error is a result of his failure to realize that none of the conditions imposed on the bath holds when we replace the bath by the universe.

The universe is not at a constant temperature. We do not know whether the universe is isolated or not. The entropy of the universe has never been defined (I am not sure that it will ever be defined), and the Helmholtz energy change of the system is *not* a disguised form of entropy change of the universe (whatever this means – nothing!).

We have stated the principle of the minimum of the Helmholtz energy with respect to all possible constrained equilibrium states having the same (T, V, N).

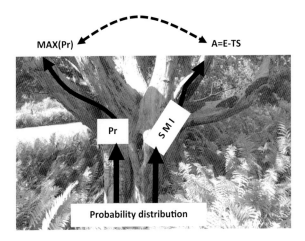

Figure 4.12 A schematic illustration of the Helmholtz energy formulation of the Second Law, and its relationship with the probability formulation.

As we have discussed in connection with the entropy formulation, one can also formulate here a principle of extremum with respect to the probability distributions of locations and momenta. Look at Fig. 4.12 and compare it with Fig. 4.9.

As in Section 1.6, we can define the SMI on any distribution (here, we restrict ourselves to distribution of locations and momenta). This SMI will change with time once we release any constraint in the system. There is one distribution which maximizes the SMI. The same distribution also maximizes the super probability, and we refer to this distribution as the equilibrium distribution. At equilibrium the SMI attains its maximum value, and it becomes proportional to the entropy of the system. This in turn defines the minimum of the combination of $E - T \times SMI$ which becomes the Helmholtz energy of the system.

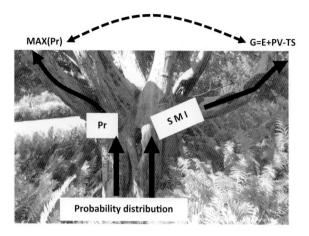

Figure 4.13 A schematic illustration of the Gibbs energy formulation of the Second Law, and its relationship with the probability formulation.

Note again that the *cause* of any process is ultimately the kinetic energy of the particles. The direction of the change is biased by the super-probability. For large systems the super-probability is overwhelmingly larger for the final equilibrium state compared with all other initial states. This explains the seemingly one-way or unidirectional change towards equilibrium. As a *result* of this change the Helmholtz energy decreases (provided we started with a well-defined constrained equilibrium state). Thus, the negative change in A is a *result*, not the *cause* of the spontaneous process occurring at constant (T, V, N).

Pause and ponder.

Please read the questions I suggested at the end of Section 4.4. Try to answer these questions using the Helmholtz energy formulation of the Second Law.

4.7 THE GIBBS ENERGY FORMULATION OF THE SECOND LAW

In this section, we shall briefly repeat the same arguments as in Section 4.6 to derive the Gibbs energy formulation of the Second Law. This formulation is valid for the processes carried out at constant temperature (T), and pressure (P). The system is still closed, i.e. N (or $N_1, ..., N_c$) is constant.

Before we formulate the relevant SL for such a system, we define the Gibbs energy[14] by

$$G = E - TS + PV$$

As in Section 4.6 we note here that the definition of the Gibbs energy applies to any thermodynamic system at equilibrium. We have the liberty to choose the independent variables characterizing the system.

However, for the Gibbs energy formulation of the SL we must choose the specific independent variables (T, P, N). Here is the Gibbs energy formulation of the SL for the T, P, N system:

For any (T, P, N) system at equilibrium the Gibbs energy has a minimum over all the possible constrained equilibrium states of the same system.

It is important to emphasize that this formulation of the SL is valid for a system at constant temperature (isothermal), constant pressure (isobaric), and closed (i.e. impermeable to particles).

An equivalent statement of the Gibbs energy formulation is:

Removing any constraint from a constrained equilibrium state of a (T, P, N) system will result in a decrease in the Gibbs energy.

As we emphasized in Section 4.6, the Gibbs energy is defined for a well-defined thermodynamic system at equilibrium. It is not true that the Gibbs energy decreases in any process occurring in any system.

The Gibbs energy formulation can be derived either from the entropy formulation, or from the Helmholtz energy formulation. Mathematically, the proof involves the Legendre transformation. One can start from $S(E, V, N)$ then change two variables; from E to T, and from V to P. Defining the Gibbs energy as a function of variables (T, P, N), one can derive the Gibbs energy formulation. Alternatively, one can start from the Helmholtz energy $A(T, V, N)$ and change one variable; from V to P, and define the Gibbs energy as a function of the new variables (T, P, N), i.e.

$$G(T, P, N) = A(T, P, N) + PV(T, P, N)$$

Note that we use here the independent variables T, P, N for all the quantities G, A, and V. Mathematically, one can prove that from the existence of minimum in the Helmholtz energy (see Section 4.6), it follows that $G(T, P, N)$ has minimum as stated above.

One can also give an "experimental" argument similar to the one provided in Section 4.6. To do this, consider a system in a heat bath and a pressure bath. In other words, the system can exchange heat with bath, as well as volume (i.e. movable or flexible walls surrounding the system). The bath is large enough so that whatever exchange of heat or volume between the system and the bath, the temperature and the pressure of the bath will not be affected. The system plus the bath, referred to as the *total*, is presumed to be isolated.

Exercise: Show that from the entropy formulation of the SL applied to the *total*, the Gibbs energy formulation of the SL for (T, P, N) system follows. We leave the details and an exercise for the reader. (See Note 15).

After doing this exercise, I suggest that you read pages 89 to 91 in Atkins' book, T4L. You will find there the same error done in connection with ΔA. $\Delta G(sys)$ is related to the $-T\Delta S(universe)$, forgetting that the universe is not a bath at constant temperature and constant pressure, and forgetting that we do not know whether the universe is an isolated system or not.

On page 90 of Atkins' book, we find:

*"There is another 'just as' to note. Just as a change in the Helmholtz energy is a **disguised** expression for the change in total entropy of the universe when a process takes place at constant volume…"*

"…so the change in Gibbs energy can be identified with a change in total entropy for processes that occur at constant pressure: $\Delta G = -T\Delta S(total)$."

This is yet another misleading and meaningless statement! It is sad to note that a book which is addressed to the lay reader presents such a distorted view of entropy, Helmholtz energy and Gibbs energy.

Finally, we draw a schematic relationship between the super probability and the Gibbs energy for the T, P, N system. As we have done in the previous section, we start with a system characterized by (T, P, N). On this system we can define the SMI on any distribution (of locations and momenta of all particles). We can also define the super probability of that distribution.

When we remove a constraint in a (T, P, N) system, the entropy change can either be positive or negative, but the ΔG must be negative. Again, we note without proof that the direction of the change in G is determined by the overwhelmingly larger probability of the final equilibrium state. As a *result* (not the *cause*) of this change, the Gibbs energy will decrease.

Pause and ponder.

Please read the questions I asked at the end of Section 4.4. Try to answer these questions using the Gibbs energy formulation of the Second Law.

4.8 APPLICATIONS OF THE SECOND LAW

The Second Law is extremely useful in many problems in physics, chemistry, biochemistry and more. These are discussed in most textbooks on thermodynamics [see also Ben-Naim (2016)].

It should be mentioned that in real applications, the Gibbs energy and the Helmholtz energy formulations of the Second Law are far more important than the entropy formulation. For instance, the existence of chemical equilibrium and the corresponding equilibrium constant is usually discussed within the Gibbs energy formulation, i.e. the principle of minimum Gibbs energy in a system at

(T, P, N) constant. Although we shall not discuss the *chemical* potential in this book (it was mentioned in Section 2.3. In connection with material equilibrium), it should be said that this is one of the most important quantities featuring in any discussion of chemical equilibrium.

In this section, we describe two applications in a very qualitative manner. In both examples we will show the difference between the extremum with respect to an arbitrary distribution on one hand, and an extremum over all constrained equilibrium states, on the other hand. In both cases, we can "invent" a catalyst (or an inhibitor) which can "freeze in" a molecular distribution so that the system becomes a constrained equilibrium system. The discussion in the following sections is very qualitative, for more quantitative treatment, the reader is referred to the literature.

4.8.1 A System Having N Solvent Molecules and One Simple Solute Particle s

We start with a system of N molecules; say water, at equilibrium, described by (T, P, N). This is a well-defined thermodynamic system at equilibrium. We can define the entropy, Helmholtz energy, or Gibbs energy for this system.

Next, we add a single solute particle s at a *fixed* position R_s, having no velocity. In practice, we cannot do this, but in theory we can think of a system of N water molecules subjected to a field of force equivalent to that produced by the solute at the fixed location R_s. For this system we can also define the entropy, the Helmholtz energy, or the Gibbs energy. Clearly, the fixed location of a solute may be viewed as a molecular distribution of locations of the solute (in this case the distribution is a Dirac delta function).

We now view this system as a *constrained* equilibrium system. We can think of hypothetical "glue" which fixes the location of the solute particle, so that it is prevented from leaving that point, in spite of the constant motion of the water molecules that surround it. In this view we have a constrained equilibrium state.

Now we remove this hypothetical "glue," i.e. remove the constraint on the fixed location and velocity of the solute, (Fig. 4.14). What will happen, and why will it happen?

Clearly, the initial state of the system is a constrained equilibrium state. The constraint is the fixed position and velocity of the solute particle. To this system, we can define the Gibbs energy which we denote by $G(T, P, N; R_s)$ which means the Gibbs energy at (T, P, N) with an additional solute particle at R_s. After the release of the constraint the solute particle will move around and, in

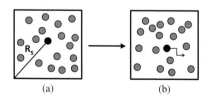

Figure 4.14 (a) A solute particle, s at a fixed position in a solvent. (b) The solute is released to access any point in the system and attains a Maxwell-Boltzmann velocity distribution.

the absence of gravitational field, after a short time the probability distribution of the locations of the solute particle will be uniform, and its velocity distribution will be Maxwell-Boltzmann (i.e. Normal distribution) along the three axes x, y, z.[16]

The distribution of locations and velocities (or momenta) of the solute at equilibrium is the one which *minimizes* the Gibbs energy of the system. Also, the super-probability will have a maximum. We started with a probability distribution which is a Dirac-delta function for both the location and the velocity (i.e. a very sharp probability to be at R_s and having velocity $v_s = 0$). We ended with a uniform distribution for the locations and Maxwell–Boltzmann distribution of the velocities. This final distribution has the maximum probability (there is an extremely small probability of finding the system such that the solute will be at R_s and having zero velocity).

Clearly, the reason for evolving from the initial to the final distribution is probabilistic. As a result of this process the Gibbs energy of the system will decrease, and we can calculate the total amount of this decrease in Gibbs energy.[17]

4.8.2 A System with *N* Solvent Molecules and One Solute Particle *s* Having One Internal Rotational Degree of Freedom

The second example is less trivial than the example discussed in Section 4.8.1. It is also more important, and it reveals a serious pitfall in protein folding theory

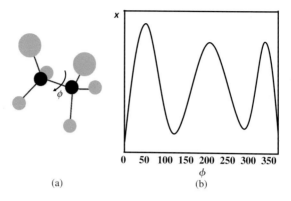

Figure 4.15 (a) A derivative of ethane and (b) a schematic possible probability density as a function of the angle ϕ.

[(Ben-Naim (2016))]. The reader who is not familiar with the protein folding problem can skip this subsection.

Consider again a system at (T, P, N, N_s), where N is the number of solvent molecules, say water, and N_s is the number of solute molecules, say a derivative of ethane, (Fig. 4.15a). The system is at equilibrium and therefore the Gibbs energy of the system is defined.

The solute molecule is a derivative of ethane; it has one internal rotational degree of freedom, (Fig. 4.15a). At equilibrium the angle, ϕ can attain any value $0 \leq \phi \leq 2\pi$ with probability distribution, $x(\phi)$ as shown schematically in Fig. 4.15b. Note that for ethane the probability distribution would have three equal minima and three maxima. When some of the hydrogens are replaced by other atoms, say chlorine or iodine, the probability distribution might still have three minima and maxima, but the values of the probability density are different at the three minima and at the three maxima.

For simplicity, suppose the angle ϕ can attain only a finite number of values between zero and 2π, say 10 angles

$$0, \frac{2\pi}{10}, \frac{2\times 2\pi}{10}, \frac{3\times 2\pi}{10}, \cdots, \frac{9\times 2\pi}{10}$$

We shall refer to a molecule with angle ϕ_i as the *i*th conformer. Clearly, an arbitrary distribution of conformers $(N_0, N_2, ..., N_9)$ is a molecular distribution, but in general, is not an equilibrium state.

We now "invent," or perhaps *imagine* a catalyst such that transition between the various conformers can occur only in the presence of a tiny amount of the catalyst. We assume that the catalyst itself is very small and does not have any other effect on the properties of the system. Thus, in the presence of the catalysts, the distribution of solute molecules in the various conformers will be the equilibrium distribution of conformers. If we remove the catalysts then we could prepare the system having any arbitrary conformer distribution $N_0, N_1, ..., N_9$ (such that $\sum_{i=0}^{9} N_i = N_s$). Such a system may be viewed as a constrained equilibrium state of the system described by (T, P, N, N_s). The constraint here is the absence of the catalyst. Insertion of the catalyst is equivalent to removal of the partitions in the example in Fig. 4.5. Alternatively, we could imagine that instead of a catalyst we have an inhibitor for the conversion between the conformers. In this case, *adding* the inhibitor is tantamount to imposing a constraint, and removing the inhibitor is equivalent to removing the constraint.

With the help of the catalyst (or the inhibitor) we can explain the difference between an arbitrary distribution of conformers and a constrained equilibrium distribution of conformers.

Suppose we prepare a (T, P, N, N_s) system with an arbitrary distribution of conformers $N_0, N_1, ..., N_9$. Corresponding to this distribution we can define the probability distribution $(x_0, x_1, ..., x_9)$ with $x_i = N_i/N_s$ and $\sum_{i=0}^{9} x_i = 1$. This probability distribution does not have to be an equilibrium distribution. It might change with time towards the equilibrium distribution. However, for any arbitrary distribution we can define the SMI, as well as the super probability on this distribution, and seek for the maximum value of the super probability which is attained at equilibrium.

All this can be done in the presence of the catalyst (or absence of inhibitor). However, as long as the system does not reach an equilibrium state we cannot define the entropy of the system, nor the Gibbs energy of the system. At this point, we use the trick of removing the catalyst (or introducing the inhibitor). Now, the system can be said to be in a *constrained* equilibrium state. The initial distribution $x^{(in)} = (x_0, x_1, ..., x_9)$ is fixed, the total system is at equilibrium (much as the system in Fig. 4.5 with the partitions), and we can define the entropy, as well as the Gibbs energy of the system. To be more precise, for any arbitrary distribution of conformers in the absence of a catalyst (or in the presence of inhibitors), we can define the Gibbs energy function $G = G(T, P, N; x^{(in)})$. Note again that the letter G is used both for the *value* of the Gibbs energy and the *name* of the function. This is well-defined for any initial distribution which corresponds to a constrained equilibrium state. Next, we release the constraint on x. We can either do it by adding a catalyst or removing an inhibitor. What will happen and why?

The answer to the first question is immediate. The system will evolve towards a new equilibrium state at which the distribution is $x^{(eq)}$. The reason the system evolves to, and once it reaches stays at the equilibrium state is because this state has the largest super probability, i.e. $\Pr(x^{(eq)})$ is maximum.

One can also prove mathematically that the passage from the initial distribution $x^{(in)}$ to the final $x^{(eq)}$, the Gibbs energy of the system will decrease [for a formal proof, see Ben-Naim (2016)]. We also note that the equilibrium distribution $x^{(eq)}$ is the distribution which maximizes the super probability Pr, as well as minimizes the Gibbs energy.

We can now repeat the same argument for any number of conformations and reach the same conclusion. In the limit we have a continuous distribution x, which is a whole function, the components of which are $x(\phi)$, such that $x(\phi)d\phi$ is the probability of finding the solute molecule at a conformation between ϕ and $\phi + d\phi$. This is the distribution we normally plot for such a molecule, as we showed in Fig. 4.15b. To any distribution $x(\phi)$ one can define the Gibbs energy *functional* $G = G(T, P, N; \boldsymbol{x})$. Again, we recall that we must assume that this distribution can be obtained by a constrained equilibrium state. Note that here the left G stands for the *value* of the Gibbs energy, while the G on the right hand side is the *name* of the functional; a functional is a function of a function, i.e. it is a function of the entire distribution function $x(\phi)$ which we denoted by $\boldsymbol{x}$. Note that the function $G(T, P, N; \phi)$

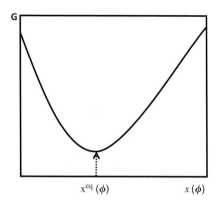

Figure 4.16 A schematic depiction of single minimum of the Gibbs energy functional $G(T, P, N; x(\phi))$. Note that in the abscissa in Figure 4.15b we have the different angles ϕ, here the abscissa represents all possible distribution functions.

may have several minima at several angles ϕ, on the other hand the Gibbs energy functional $G(T, P, N; x)$ has a single minimum at a specific distribution function $x(\phi)$. This is shown schematically in Fig. 4.16.

The Gibbs energy formulation of the SL for this system is as follows:

Starting with any initial distribution $x^{(in)}$, and releasing the constraint on this distribution (at constants T, P, N, N_s), the system will evolve, with high probability to a new equilibrium distribution $x^{(eq)}$ such that the Gibbs *functional* $G(T, P, N; x)$ attains its minimum with respect to all possible distributions.

One can easily generalize to any molecule with any number of internal rotational angles. Proteins are examples. It is here that a serious pitfall exists into which many have fallen. We describe this pitfall here very briefly. For a more complete discussion, [see Ben-Naim (2015, 2016)].

Let ϕ be the set of all the rotational angles which describes the conformation of a protein. A protein of some 150 amino acids has at least 300 angles that define its conformation.

Suppose we have one protein in a solvent characterized by (T, P, N) which are fixed. We can ask two questions about this protein:

1. What is the probability of finding the protein at a particular conformation ϕ?
2. What is the probability of finding a particular distribution of the conformations, say, $x(\phi)$?

The Gibbs energy formulation of the SL states that, at equilibrium the Gibbs energy *functional* $G(T, P, N; x(\phi))$ is minimum overall possible distributions $x(\phi)$.

Unfortunately, most people working on proteins have searched for the *global minimum* of the Gibbs energy *function* $G(T, P, N; \phi)$, referred to as the *Gibbs energy landscape*, hoping to find the native conformation of the protein at this minimum.

Of course, the native conformation of the protein might, or might not be at the global minimum of the Gibbs energy landscape — but this has nothing to do with the Second Law of Thermodynamics.

As we have noted in earlier, the system's state can fluctuate around the equilibrium state. However, the Gibbs energy of the system is the minimal value of its

Gibbs energy functional. As such its value is fixed, and does not fluctuate with time.

4.9 SOME FINAL THOUGHTS ON THE SECOND LAW

The Second Law was enunciated and used for over a hundred years without any reference to the atomic nature of matter. Also, the concept of entropy was defined and used successfully by many scientists without bothering about its meaning, or about the question of whether or not it has a meaning. The entropy was intimately associated with the Second Law; in fact, it was the quantity with which the Second Law was formulated.

In science, one usually starts from a series of experiments or observations, and if a persistent pattern is observed, one enunciates a law of physics, chemistry, or biology. One cannot prove the veracity of that law. For as long as it is deemed consistent with new experiments, the law is considered to be valid.

This procedure is different from an enunciation of a mathematical theorem. You can measure the length of the three sides of a right triangle, and find that the square of the hypotenuse is equal to the sum of the square of the two other sides. You can repeat these experiments hundreds, thousands, or even millions of times, and still find the same equality. It is pronounced a theorem only when one can prove it.

In physics, one cannot prove the validity of a law. It is assumed to be true until it fails. The conservation of energy and the conservation of matter were two distinct laws believed to be absolutely true, until it was found that matter can be converted to energy, and energy to matter. The two laws failed, and a new law of conservation of mass-energy had to be pronounced.

The situation with the Second Law is much more complicated. As long as the meaning of entropy was not understood, the Second Law was applied to all systems, animated, inanimate, and even the entire universe. The entropy was viewed as the *cause* of every process, everything that changes, every thought, every feeling, every creation of art, or even social processes. This is the main reason why the entropy and the Second Law were considered the most mysterious concept in Physics. Many scientists considered entropy to be forever a mystery. For more details, quotations and reasons for the mystery associated with entropy, see Ben-Naim (2016c).

Once the meaning of entropy is understood, the mystery associated with entropy and the Second Law evaporates. As I have discussed in my previous books, Ben-Naim (2008, 2016c), the Second Law may be viewed as a *law of probability*, rather than as a law of physics. The most important outcome of the understanding of entropy is that entropy is not the *cause* of anything that happens, whether in a well-defined experiment of mixing of two gases, or the birth and the death of a living system. Once we know what entropy is, we also know for which systems it can be defined. The consequence of such knowledge might come as a shock to those who apply the concept of entropy to living systems and to the entire universe.

It is shocking to find out that the entropy — thought to be the *cause* of everything that happens, that drives everything — ravages everything, is in fact completely innocent, impotent, and drives nothing!

Traditionally, the Second Law was formulated in terms of the entropy. We presented a more general formulation of the Second Law based on probability. The entropy is a quantity that *describes* or characterizes the system at equilibrium. The Second Law deals with the question of why a system will tend to equilibrium — it is only in special isolated systems (and the least interesting ones) that entropy increases. Both entropy and the Second Law are based on probabilities. But these probabilities are different. Entropy, as a special case of a SMI is a *pure probability* quantity. It is defined on the PD of observing a certain configurations (locations and velocities) of all particles. The Second Law is concerned with the super probability of observing a probability distribution. This super probability is related to the entropy only for isolated systems. Since in practice we never deal with isolated systems, the entropy formulation of the SL is the least useful one. On the other hand, the same super probability which governs any evolution from one equilibrium state to another is related to either the Helmholtz, or the Gibbs energy for a (T, V, N), or a (T, P, N) system, respectively. These are the most common systems used in the laboratory and therefore the latter formulations of the SL are more important.

In Fig. 4.17, we summarize the four formulations of the Second Law. At the very root of the definition of both the SMI and the super probability lies a *molecular*

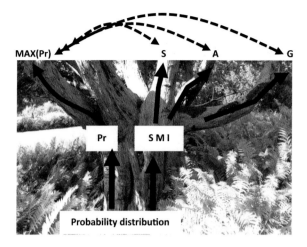

Figure 4.17 Schematic illustration of the four formulations of the Second Law, and its relationship with the probability formulation.

probability distribution. We have discussed qualitatively only ideal gases of simple particles. For such cases, we need only the PD of locations and velocities (or momenta). In more general cases, we must also take the distribution of particles in various energy levels corresponding to internal degrees of freedom (electronic, vibrational, rotational, etc.).

On any PD one can define two *functionals*; the SMI and the super probability. Both are defined for any probability distribution, and for any system (small or large, at, or not at equilibrium). Once we remove the constraints, the probability distribution might change. Also, the two functionals: SMI and Pr might change as a result of the evolution of the system towards the new equilibrium state.

Once an equilibrium state is reached, the super probability reaches a maximum. At that state, the value of the SMI is proportional to the entropy of the system. It is only at this state of equilibrium that one can formulate the Second Law. The most general formulation is in terms of Pr; it applies to any thermodynamic system. For specific systems, such as (E, V, N), (T, V, N) or (T, P, N), the entropy, Helmholtz energy, and the Gibbs energy formulations apply, respectively. Figure 4.17 shows only the three most common thermodynamic systems. There are many other characterizations of thermodynamic systems for which one can define a thermodynamic quantity, and which attains an extremum at equilibrium. For all of these the probability formulation always hold true.

One final comment regarding the conditions under which we define the thermodynamic quantities is in order. Figure 4.18a shows an initial constrained equilibrium state. We release the constraints and the system evolves to a new (unconstrained) equilibrium state, Fig. 4.18b. The entropy (as well as any other thermodynamic function) is *defined* only at the equilibrium states; *before* and *after* the removal of the partitions.

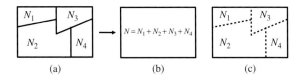

Figure 4.18 (a) A constrained equilibrium system. (b) The unconstrained equilibrium state. (c) The same system as in (b) (unconstrained) but with a distribution of particles as in (a).

On the other hand, if we want to compare the super probabilities of the initial and the final states, we do that *after* the removal of the constraint. Figure 4.18c shows the system at the final equilibrium state. At this state we can ask about the probability (to which we referred as super probability) of observing the *initial* distribution, say, of location, in the system when there are no partitions (shown in Fig. 4.18c by dashed lines). This initial distribution is the same as the one we had *before* the removal of the constraint. However, the super probability of occurrence of this distribution refers only to the system *after* the removal of the constraint.

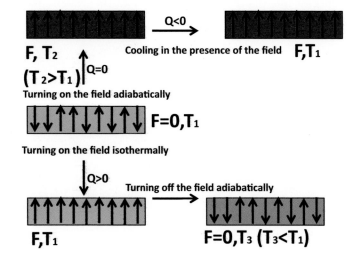

Various Processes in Magnetic Cooling

5

The Third Law (TL)

The Third Law (TL) has its origin in chemical transformation. It was originally referred to as: The *Heat Theorem*, or the Nernst heat theorem. It states:

The **entropy change** *in any isothermal reaction tends to zero when* $T \to 0$.

$$\lim_{T \to 0} \Delta S = 0$$

Examples of processes, where the observed entropy changes tend to zero at $T \to 0$ are:

1. $\Delta S = S$ (*Monoclinic sulphur*) $- S$ (*Rhombic sulphur*) $\approx 0.02\,eu$
2. $\Delta S = S$ ($\alpha - phosphine$) $- S$ ($\beta - phosphine$) $\approx 0.01\,eu$

3. The phase transition between liquid and solid helium
4. A chemical reaction such as

$$Pb + I_2 \rightleftarrows PbI_2$$

This theorem has been transformed and reformulated several times, and now it is known as the Third Law of Thermodynamics. In fact, none of these formulations was found to have a general validity. Personally, I doubt that this theorem deserves to be referred to as a Law of Thermodynamics. It is brought up in this book mainly because such a Law appears in most textbooks as one of the Four Laws of Thermodynamics. The second reason is, that I committed myself in the title of the book to discuss the Four Laws that do not Drive the Universe. Needless to say, this Law drives nothing!

I would like to remind the readers that Clausius did not define the entropy of a system, but a change of entropy associated with a specific process. Boltzmann seems to have defined an absolute entropy with his famous formula ($S = k\log W$). The latter was confirmed as being "identical" to Clausius' definition only when changes in entropy calculated by the Boltzmann formula agreed with those calculated, or measured by Clausius' definition. The same is true for the definition based on SMI which was discussed in Section 4.5.3.

5.1 THE VARIOUS FORMULATION OF THE TL

The Nernst heat theorem deals with the limit of the entropy change in an isothermal process. If one draws

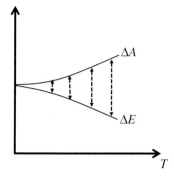

Figure 5.1 The change in ΔE and ΔA as we approach low temperatures.

the Gibbs energy, or Helmholtz energy change and the internal energy change as a function of time one gets curves like those shown in Fig. 5.1.

Thus, we see that as the temperature decreases, the difference between ΔA and ΔE seems to tend to zero. The data for all such experiments are always for positive, non-zero temperatures.

The other formulation is:

The entropy of any pure stable crystalline compound tends to zero at $T = 0K$.

This formulation is sometimes presented as a postulate [see Callen (1985)]. It is clearly a stronger statement of the Third Law compared with Nernst's.

Finally, we mention another formulation of the TL which is enunciated as follows:

It is impossible by any procedure, no matter how idealized, to reduce the temperature of a system to the absolute zero in a finite number of operations.

For more details on the various formulations of the TL, see Fowler and Guggenheim (1956).

5.2 APPLICATIONS OF THE THIRD LAW

There are not many applications of the TL. In fact, one could do all thermodynamic calculations or experiments without even mentioning the TL.

One of the most interesting concepts associated with the TL is the so-called *residual entropy*.

Recall that Clausius' definition of entropy, as well as most of the applications of entropy in thermodynamics involve the difference between the entropy values of the system at two different states. On the other hand, Boltzmann's method, as well as the method based on the SMI provides "absolute" values of the entropy.

We enclosed "absolute" in quotation marks because, up to this point, we cannot verify that the calculated theoretical values obtained from the entropy function are the "true" values. How can we compare the calculated theoretical values of entropy with experimental results? Here, the third law of thermodynamics comes to our aid. The third law basically states that the entropy of a system tends to zero when we approach the absolute zero temperature ($T = 0K$). Whenever this is true, one could in principle calculate from experimental data (on heat capacity, heat of melting, heat of evaporation, etc.) the *difference* between the entropy of the actual system and the entropy of the system at $T = 0K$. If the entropy of the system at $T = 0K$ is indeed zero, then the difference $S(T) - S(T = 0)$ will be the "absolute" value of $S(T)$.

Table 5.1 Calculated and Measured Entropies of Gases*

Substance	Temperature	Entropy (E.U.)	
		Theoretical	Experimental
A	B.P.	30.87	30.85
O_2	B.P.	40.68	40.70
N_2	B.P.	36.42	36.53
Cl_2	B.P.	51.55	51.56
HCl	B.P.	41.45	41.3
CH_4	B.P.	36.61	36.53
C_2H_4	B.P.	47.35	47.36
NH_3		44.1	44.1
CO_2		47.5	47.5
CH_3Br		58.0	57.9
CH_3Cl	298K	55.9	55.9

*B.P is the boiling temperature. E. U. are entropy units of cal/mol. K

In fact, when we calculate the entropy of some gases based on experimental data and compare it with the entropy of the same system based on the theoretical entropy function, we get very good agreement between the two results. In such cases, we can remove the quotation marks and refer to the entropy as the *absolute* entropy of the gas.

Table 5.1 shows values of the entropy of a few gases calculated from experimental data, and from the theoretical equation for the entropy.

The agreement between the theoretical and the experimental values is, in fact, proof of the success of the Boltzmann entropy, as well as the entropy based on the SMI.

There are however, many cases where there is no agreement between the theoretical and the experimental

Table 5.2 Calculated and Measured Entropies of Gases*

		Entropy (E.U.)		
Substance	Temperature	Theoretical	Experimental	Residual Entropy
CO	B.P.	38.32	37.2	1.22
NO	B.P.	43.75	43.03	0.72
N_2O	B.P.	48.50	47.36	1.14
H_2O	298K	45.10	44.29	0.81
D_2O	298K	46.66	45.89	0.77
CH_3D	B.P.	39.49	36.72	2.77

*E.U. are entropy units, see Table 5.1

values. We denote by S_{exp} the experimental value, and by S_{theor} the theoretical value of the entropy.

Table 5.2 shows a few examples where there is a discrepancy between S_{exp} and S_{theor}. We define residual entropy by the difference between the experimental and the theoretical values

$$\Delta S = S_{theor} - S_{exp}$$

There are many reasons for discrepancies between the theoretical and experimental values of the entropy. There could be inaccuracies in the values of the heat capacities, or the heat of melting, or the heat of boiling. [For details, see Wilks (1961)]. However, there are some discrepancies which can well be explained by the so-called configurational degeneracy of the crystals at very low temperatures.

Consider the following simplest examples; linear molecules such as HCl, HBr, or HI. When the crystals of these molecules are cooled down to very low temperature, there is only one arrangement of all the molecules in the crystal. The reason is that if you change the orientation, say of HCl from HCl to ClH, the energy of the crystal will change considerably. Therefore, as the temperature is lowered, the molecules will tend to orient themselves in such a way that the total energy of the system will be the lowest. In such cases it is believed that there is only one configuration of the molecules at 0K, hence $S(0) = k_B \ln 1 = 0$.

On the other hand, when you cool crystals of molecules such as CO, NO, and NNO, there is not much difference in the interaction energy between the molecules having either the CO, or OC configurations. Figure 5.2 shows a possible configuration of a crystal CO. If this assumption is correct, then we can calculate the total number of configurations in a crystal of one mole of CO. Assuming that each molecule can be in either one of the two orientations (CO or OC), then there are altogether 2^N possible configurations. Assuming that the energy of all these configurations are nearly equal, we can calculate the Boltzmann entropy for this configurational degeneracy, which is

$$k_B \ln 2^N \approx 1.38 \ cal/molK$$

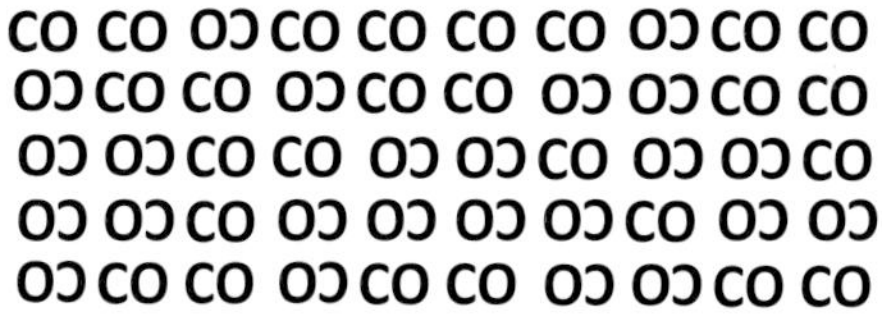

CO CO OC CO CO CO CO OC CO CO
OC CO CO OC CO CO OC OC CO CO
OC OC CO CO OC OC CO OC OC CO
OC OC CO OC OC OC OC CO OC CO
OC CO CO OC CO CO OC OC CO CO

Figure 5.2 A possible configuration of a crystal of CO. It is assumed that the energy of the crystal almost does not change when the orientation of the molecule changes from CO to OC.

This is very close to the residual entropy of the molecules such as CO, NO, and NNO as shown in Table 5.2.

Another interesting example is the case of methane (CH_4), methyl chloride (CH_3Cl), and methyl bromide (CH_3Br). In all of these molecules the residual entropy is nearly zero. On the other hand, if we take the molecule CH_3D (i.e. methane in which one hydrogen is replaced by deuterium), we find considerable residual entropy. It can be shown that each molecule of CH_3D can be in one of four configurations, so that for N molecules we will have 4^N possible configurations. Assuming (approximately) that all these configurations have the same energy (therefore, equally probable) we arrive at the estimate of the residual entropy of CH_3D which is, for one mole of CH_3D

$$k_B \ln 4^N \approx 2.75 \; cal/molK$$

which is very close to the value shown in Table 5.2.

We can conclude that if there are many configurations which have nearly the same energy, the system will have a residual entropy which may be calculated from Boltzmann's equation $S = k_B \ln W$.

Perhaps the most interesting case is that of the residual entropy of ice. This is an important example demonstrating how a simple statistical calculation led to understanding the concept of the *residual entropy of ice*, and indirectly contributed to the understanding of *entropy* in general. This calculation was first published by Linus Pauling in 1935.

The residual entropy of ice is a measure of the number of configurations, or the number of arrangements of the hydrogen atoms on the fixed lattice of ice. A detailed discussion of Pauling calculation can be found in Ben-Naim (2016c).

5.3 ADIABATIC COOLING

In most textbooks discussing the TL of thermodynamics you will also find a discussion of adiabatic de-magnetization. The reason is that the TL involves the question of the attainability of the absolute zero (see Section 5.2). Therefore, it is natural to ask, how do we get to very low temperatures experimentally?

The method of adiabatic cooling is interesting in its own right even without any reference to the TL. It actually uses the First, and the Second Laws, not the TL.

Before we discuss adiabatic de-magnetization we discuss two processes of adiabatic expansion in which we achieve cooling. All these examples have one principle in common. We perform a spontaneous process which causes an increase in potential energy. Since the process is done adiabatically, this increase in potential energy must come from *reduction* in the kinetic energy, hence the decrease in temperature.

(a) Adiabatic Expansion of Non-Ideal Gas

Consider a gas or a fluid at some temperature $T^{(in)}$ and volume $V^{(in)}$. The atoms or molecules in the fluid are at such distances that on average they attract each other. A schematic pair potential between two argon atoms is shown in Fig. 5.3.

Now suppose that a partition is removed from the system separating two compartments as shown in Fig. 5.4a.

When the gas expands from V to $2V$ the average distance between the particles will increase. The interaction energy among all the particles will, on average, decrease.

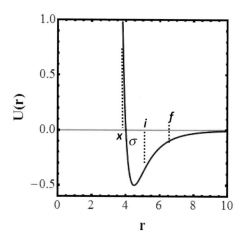

Figure 5.3 The general form of a pair potential between two particles. On the right of the minimum the two particles attract each other, on the left of the minimum the two particles repel each other.

Since the process was done adiabatically, $Q = 0$ (no exchange of heat with the surroundings), and also $W = 0$ (no exchange of work with the surroundings). Therefore, from the FL, we must have $\Delta E = 0$, i.e. no change in the internal energy of the system.

In non-ideal systems the internal energy is the sum of the total interaction energy among the particles, and the kinetic energy of all the particles. In this particular expansion process (Fig. 5.4a) the potential energy increases (on average) from (*i*) to (*f*) in Fig. 5.3. Therefore, the kinetic energy must *decrease* to maintain a constant total internal energy. This *decrease* in the average kinetic energy is the same as lowering the

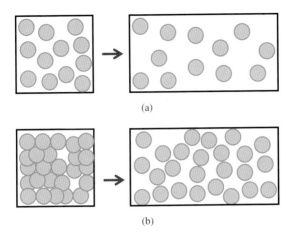

Figure 5.4 Adiabatic expansion of a system of interacting particles from V to $2V$. (a) Expansion associated with cooling, (b) expansion associated with heating.

temperature. This procedure is essentially the one that we use in refrigerators.

<u>Exercise</u>: Suppose we start with highly condensed argon atoms, (Fig. 5.4b), such that the initial distance between pair of particles is $r < \sigma$ as indicated by x in Fig. 5.3. We now expand from V to $2V$ in an adiabatic process. How will the temperature of the system change in this process[19]?

(b) Adiabatic Expansion of Ideal Gas in a Gravitational Field

We next discuss a process of cooling which brings us closer to the adiabatic de-magnetization process.

Consider an enclosed column of gas in a gravitational field occupying a volume V, and having an average

146   *The Four Laws That Do Not Drive The Universe*

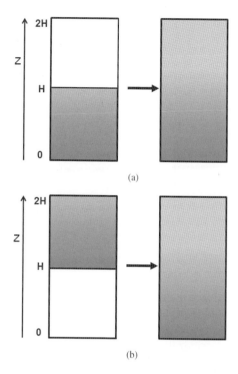

Figure 5.5 Adiabatic expansion of an ideal gas in a gravitational field. (a) Expansion with cooling, (b) expansion with heating.

temperature $T^{(in)}$, in Fig. 5.5a. Unlike the previous example, we have an ideal gas so that intermolecular interactions are negligible. In this case there is a potential energy for each molecule which depends on the height of the molecule. This gravitational potential energy will cause a non-uniform distribution of the density along the height of the column which is known as the barometric, or the barometric exponential distribution.

In the initial state, let us denote the average potential energy of the molecules by $mg\langle z\rangle^{(in)}$ where m is the mass, g the gravitational acceleration, and $\langle z\rangle^{(in)}$ the average height of the molecules. We now remove the partition and let the system expand adiabatically between V and $2V$ (the maximum height changes from H to $2H$). What will happen?

As in the previous example, the expansion is adiabatic and no work is exchanged with the surroundings. Hence, $Q = W = 0$ and therefore also $\Delta E = 0$. However, the average height of the molecules will increase from $\langle z\rangle^{(in)}$, to say $\langle z\rangle^f$. This means that the potential energy will *increase*, since the total internal energy is unchanged the kinetic energy will decrease — which is the same as lowering the temperature.

All we have described above is qualitative. For more detailed elaboration on this process, see Ben-Naim (2017).

Exercise: Consider the process of adiabatic expansion of the same gas at an initial temperature $T^{(in)}$, but now expanding downwards, as in Fig. 5.5b. What would the final temperature of the gas be[20]?

(c) Adiabatic Demagnetization

This method is useful in bringing down the temperature of a system to within a small fraction of one degree above the absolute zero. This process is called magnetic cooling or adiabatic demagnetization.

The reasons for the cooling are similar to those discussed in the previous examples.

The process of adiabatic demagnetization involves the following steps:

1. A paramagnetic salt (e.g. salt of ferric ions) is first brought into a bath of liquid helium, at an initial temperature.
2. A powerful magnetic field is switched on. The magnetic dipoles orient themselves (on average) in the direction of the magnetic field. (Fig. 5.6a) (This step is the analog of compression of a gas in example (a) by doing PV work on the gas. Here, the magnetic field brings the magnetic dipoles to a lower potential energy state). If the "turning on" of the field was done adiabatically, the temperature will be raised. However, the "turning on" is done at *constant temperature*. Therefore, the initial temperature of the system ($T^{(in)}$) is maintained.
3. The system is now isolated and the magnetic field is lowered. As a result of this, there will be no preferential orientation of the magnetic dipoles, Figure 5.6b.

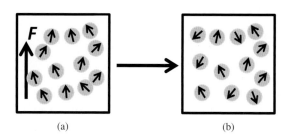

Figure 5.6 (a) Molecules having a magnetic dipole moment are preferentially oriented in a magnetic field F. (b) Removal of the field causes randomization of the orientations of the dipoles.

The average angle of orientation of the dipoles with respect to the magnetic field will be larger. This "randomization" of the dipoles will raise their potential energy. Since this process is carried out adiabatically, the increase in the potential energy of the dipoles must come from the kinetic energy of the atom. Therefore, the temperature of the salt will decrease, i.e., $T^{(f)} < T^{(in)}$.

5.4 CONCLUDING REMARKS

The Third Law can hardly be reckoned as a Law of Thermodynamics. The proof: Nothing would have changed in the world if this Law had never been discovered, never pronounced, or not even existed. In any case we are almost always interested in changes in entropies, not in their absolute values. This comment is true for the energy too. Note however, that when I say that the same is true for the energy I mean only that its absolute value is of no importance. On the other hand, it is clear that the world would have been very different — perhaps, very, very different — without the First Law, or the Law of Conservation of Mass-Energy.

As we have seen, the applications of the Third Law (if these can be called "applications" at all), are very limited. We were able to define an absolute value for the entropy, but that can hardly be reckoned as an application. The discussion of the residual entropy, and its consequence regarding the energetics of configurations of crystals, and in particular, the "structure" of ice with respect to the locations of the hydrogen atom was indeed very revealing. But that was not a result of the application of the TL. The formulation of the TL in terms of the unattainability of the absolute temperature also may not be considered as an application of the TL.

Finally, needless to say that the TL does not *drive anything* — as you might know, *anything*, includes the "universe!"

6

Which of The Four Laws Drive The Universe?

Now that I have completed the description of the "four laws," I owe you an explanation of the second part of the title of this book: "…that don't drive the universe."

By now, I am sure you realize that I have chosen my particular title for the book because it is diametrically opposed to Atkins' book: *"Four Laws that Drive the Universe."*

The first question we might ask is: "Do we know *any* Law that drives the universe?"

The answer to this question is a resounding No!

This is the answer any scientist must give you. No one knows any Law that drives the whole universe. I also doubt that we will ever know such a Law, and perhaps it is even meaningless to ask about such a Law before we can fully describe and understand the entire universe.

The second, more modest, question we might ask is: "Do we know any Law that drives all the processes that we have observed, that occur in parts of the universe?

The answer to this question is also No!

We know of different Laws that drive different specific processes; the flying of a ball, the falling of water from a high to a lower level, or the path taken by a ray of light passing through different media. But we still do not know the Laws that drive living systems in general, nor the Laws that govern our feelings, our thoughts or our motivation for creating a work of art. Of course we have no idea which Law governs social phenomena, if such a Law exists.

At this point, we can conclude that the title of Atkins' book (T4L) is outright misleading.

Since T4L, as well as the present book, describes the Four Laws of Thermodynamics, and since these Laws are associated with well-defined thermodynamic systems (not nuclear fission, not the process of reproduction of living systems, and not the expansion of entire universe), we might ask the more modest question: Which Law drives a process in a well-defined thermodynamic system?

Obviously, neither the ZL, the FL, nor the TL, *drives* any process. Therefore, this rules out the truthfulness of

75% of Atkins' title, even when it is applied to a tiny part of the universe; thermodynamic systems.

Thus, we are left with the Second Law. Does this Law drive any process in a well-defined thermodynamic system? The reader should be aware of the fact that many authors (including Atkins) of popular science books, assigning to *entropy* the power to drive many processes. Examples:

1. Waterfalls.
2. Brain activities, including thoughts, feelings, and creativity.
3. Life processes.
4. A child's room becoming messy.
5. The universe is doomed to "thermal death."

All these are typical nonsense ascribed to the almighty Entropy and the Second Law. The truth is that Entropy is not even defined for any of these processes. Therefore, entropy cannot *drive* these processes. Therefore, this rules out another 25% of Atkins's claims — which renders the title of his book 100% untrue, and 100% meaningless.

Further reducing the number of systems for which the Second Law does apply, we focus on the following three well-defined processes shown in Fig. 6.1.

We shall discuss in more details the simplest of these processes; the expansion of ideal gas, shown in Fig. 6.1a.

We start with N atoms, of say, argon in a compartment of volume V, and remove the partition. We observe that the gas *expands* into the entire new volume $2V$. What drives this process?

154 *The Four Laws That Do Not Drive The Universe*

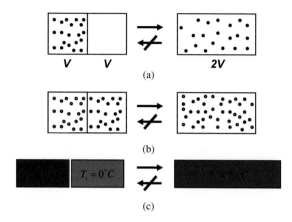

Figure 6.1 Three spontaneous processes in isolated systems for which the entropy increases. (a) Expansion of an ideal gas, (b) mixing of two ideal gases, (c) heat transfer from a cold to a hot body.

Clearly, if we did not know that the system consists of small particles like atoms and molecules, we could have never understood what causes this process.

Of course, one could follow Clausius definition of the Entropy and the formulation of the Second Law in terms of Entropy changes, and find that the Entropy has increased in the three processes described in Fig. 6.1. It is natural to conclude that the Entropy, not yet understood in molecular terms, is somehow *responsible* for, or the *cause* of, or the *driving force* in these processes. Such conclusions were in fact reached by many others, and even extended to realms for which Entropy is not even defined. As long as Entropy is not understood in molecular terms, one might suspect that Entropy is indeed the driving force for all these processes.

However, once we have grasped the meaning of Entropy on a molecular level, this conclusion becomes invalid. Entropy, by itself, does not drive any process. So the question remains: knowing what happens on the molecular level, what drives the simple expansion of the gas?

Obviously, none of the processes described in Fig. 6.1 would have occurred were the molecules devoid of kinetic energy, if all the atoms and molecules were at rest, nothing would have happened; the gas would not expand to occupy the entire volume $2V$, and the two gases would not mix in Fig. 6.1b.

Knowing the fact that without all these kinetic energies of the particles, nothing would have happened, can we say that these kinetic energies are the driving forces for the processes in Fig. 6.1? The answer is No! The particles move at random directions and at random speeds, and there is nothing in these motions to dictate the specific *direction* in which the processes in Fig. 6.1 occur.

It is at this point that one invokes the SL, or the "tendency of Entropy to increase" to explain the unique *direction* of the thermodynamic processes. Unfortunately, the uniqueness of the *direction* of the thermodynamic process is only an illusion. What we call an irreversible process (and what is claimed to be in conflict with the reversibility of the Laws of motions of the atom) is not irreversible in an absolute sense. Theoretically the system can go from occupying the entire volume $2V$, into the smaller volume V. However, it is extremely unlikely to observe such a phenomenon.

The Entropy change we observe is a *result* of the process, not the *cause* of the process.

In Section 4.4 we discussed the probability formulation of the Second Law. This formulation could not have been stated without the knowledge of the molecular meaning of Entropy. As we have found, there is a specific distribution of locations and velocities that maximizes the SMI of the thermodynamic system. The same distribution also maximizes the super probability. Thus, we can rationalize the apparently unique direction of the processes in Fig. 6.1 by saying that the system will go from a distribution (or location and velocities) having a low probability to a distribution having higher probability, and reaching the condition of maximum probability at equilibrium.

The question remains; does probability drive the process?

I claim that the direction of the expansion process is determined by the higher probability of the final state. It is not the probability that tends to increase, as Brillouin claimed, but the probabilities of all states are given, and the system spends more time in states of higher probability. This is the frequency definition of probability, and when we deal with thermodynamic systems, "more time" turns into "most time" and "most time" turns into "always."

I prefer to say that the system goes from one state to another *because* the probability of the final state is much larger than that of the initial state. If one adopts this formulation of the Second Law, one can say, more as a figure of speech, that probability is the cause for these specific processes. This is basically the Law of large numbers. It would not hold for a system of small numbers of particles [for simulated experiments with small numbers of particle see Ben-Naim (2010)]. Certainly, this Law cannot be used for explaining the driving force for our thoughts, or our feelings. As for the entire universe, let us wait for the time when we understand the behavior of the universe before guessing what drives it.

In the Preface of Atkins's book we find:

"When in due course, however, you emerge from the other end of this slim volume, with your brain more sinewy and exercised, you will have a profound understanding of the role of energy in the world. In short, you will know what drives the universe.

...the subject could touch an enormously wide range of phenomena, from the efficiency of heat engines, heat pumps, and refrigerators, taking in chemistry on the way, and reaching as far as the processes of life."

This promise begs fulfillment. After reading Atkins' book you will have no understanding of what drives the universe!

In the conclusion of Atkins's book, we find:

"What I have sought to cover are the core concepts, concepts that effectively sprang from the steam engine but reach out to embrace the unfolding of a thought. This little mighty handful of laws truly drive the universe, touching and illuminating everything we know."

Nothing can be more untrue and more misleading than the concluding sentences as quoted above.

Instead one can say with confidence that:

> ***The four laws do not drive the universe!***
> ***The four laws do not drive, nor illuminate***
> ***everything we know!***

A friend who read the first draft of the manuscript of this book commented that:

"T4L (meaning Atkins' book: *Four Laws*), did drive one thing. It drove you to write this book. In this sense, Atkins was partially right in claiming that the "Four Laws" drive something…

I agree.

Notes

Note 1: Here are some quotations from Atkins' book (2007):

"The Zeroth Law established the meaning of what perhaps is the most familiar but in fact most enigmatic of these properties: temperature."

This is a typically misleading and untrue statement. The ZL *does not* state *anything* about the *meaning* of temperature. Temperature is not, and never was an "enigmatic property." The meaning of temperature had long been established independently of the ZL.

*"First, the huge importance of the Boltzmann distribution is that it reveals the molecular significance of temperature: **temperature is the parameter that tells us the most probable distribution of populations of molecules over the available states of a system at equilibrium**."*

Indeed, Boltzmann's distribution is very important, but it does not reveal the molecular significance of temperature. Boltzmann's distribution *is* "the most probable distribution of populations of molecules…" The particular form of this distribution depends on the temperature, but it is not the temperature that "tells us the most probable distribution…"

Note 2: The transitivity of a relationship between two numbers, objects or persons, is usually taken to be either true or false on an intuitive basis. For instance, *equality* between numbers is transitive; if $A = B$ and $B = C$, then $A = C$. Also, *"greater than"* is transitive; if $A > B$, and $B > C$, then $A > C$. On the other hand, there are relationships which are not transitive. An obvious example is fatherhood (or motherhood). If A is the father of B, and B is the father of C, then it does not follow that A is the father of C. Another less intuitive example of a relationship which might, or might not be transitive is the *"supportive conditional probability."* We say that A supports B whenever the occurrence of A increases the probability of occurrence B. We write this as $P(B \mid A) > P(B)$. Intuitively, we expect that if A supports B, and B supports C, then A also supports C. This transitivity is not always true. There are examples for which this relationship is transitive, and there are examples for which it is not. For details, see Ben-Naim (2014).

Note 3: The conservation of energy, as well as of momentum applies to any other collision along a different path not necessarily along the line connecting the two centers.

The conservation of (linear) momentum is $m_1 v_1 + m_2 v_2 = m_1 v_1' + m_2 v_2'$ and the conservation of the kinetic energy in $\frac{m_1 v_1^2}{2} + \frac{m_2 v_2^2}{2} = \frac{m_1 v_1'^2}{2} + \frac{m_2 v_2'^2}{2}$ (primed letters refer to the velocities after the collision).

Note 4: The kinetic energy at time t can be written as $\frac{m[v(t)]^2}{2}$ and its potential energy as $mgh(t)$, where m is the mass, $v(t)$ the velocity at time t, g the gravitational acceleration constant and $h(t)$ the height of the weight relative to the level at its minimum, Fig. 3.2b. Assuming no friction, the sum of the kinetic and potential energy is constant.

Note 5: Work in an expansion process against a constant external pressure.
As an example we calculate the work of expansion of an ideal gas. According to the definition, the work associated with expansion of the system from volume V_1 to V_2 is

$$W = -\int_{V_1}^{V_2} P dV$$

The minus sign is added because in physics we define as positive work done *on* the system, and as *negative* when work is done *by* the system.

Suppose that some process occurs within the system, e.g., a spontaneous chemical reaction. As a result of which the piston is pushed upwards. During this process the pressure within the system will change. In general, the pressure of the system may not be defined along the process, in which case we cannot do the integration.

However, if the *external* pressure on the piston, is constant, P_{ex}, say atmospheric pressure, then we can write

$$W = -\int_{V_1}^{V_2} P_{ex} dV = -P_{ex}(V_2 - V_1)$$

Thus, since P_{ex} is defined as a positive number, and if $V_1 - V_1 > 0$, then W is negative, i.e., work is done *by* the system on its surroundings.

Normally, the pressure within the system P is not constant during the process. We wish to relate the work W to the pressure P *within* the system in the process. In order to do so we need to know how the pressure depends on the volume. We discuss two cases of such processes of expansion, for which we know the function $P(V)$, and therefore we can calculate the work done by the system in these processes.

Note 6: Work in an isothermal quasi-static expansion process of an ideal gas.

Consider a system of ideal gas enclosed in a cylinder and immersed in a thermostat, i.e., a very large bath maintained at constant temperature T.

Suppose that the system is initially characterized by the pressure P_1 and the volume V_1, and that a weight of mass M, placed on the piston, exactly balances the pressure of the gas so that the volume of the system is maintained at V_1. In the initial state the external pressure, which is equal to the internal pressure, is equal to P_1, i.e., $P_{ex} = P(M) = P_1$.

We now replace the weight of mass M by a lighter weight of mass $M/2$ exerting an external pressure $P_{ex} = P(M/2)$.

Clearly, the gas will rapidly expand, and the pressure of the gas will be reduced to say, P_2, and the new volume will be V_2. At the new equilibrium state $P_{ex} = P(M/2) = P_2$ the temperature of the gas is constant T, and knowing the equation of state of the ideal gas, we can write

$$P_1 V_1 = nRT = P_2 V_2$$

or equivalently; $\qquad \dfrac{P_2}{P_1} = \dfrac{V_1}{V_2}$

In this process, we know the initial and the final states of the gas. These two states are described by two points as shown in a PV diagram of Fig. N.1a.

The work done by the system in lifting the weight is simply:

$$W = -\int_{V_1}^{V_2} P(M/2)\,dV = -P(M/2)(V_2 - V_1)$$

By replacing the initial weight of mass M by a new weight of mass $M/2$, the external pressure on the piston has changed from $P(M)$ to $P(M/2)$, and the new pressure is maintained during the entire process.

We now wish to express the work done by the system in terms of the parameters P and V of the system. Here we encounter a problem. As we change the weight on the piston, the system expands rapidly. During this process the pressure in the system is not uniform. If we were to measure the presence at different distances from the piston we would have found that initially, as the piston moves upward the pressure just below the piston will be reduced, while the pressure at the bottom of the

164 *The Four Laws That Do Not Drive The Universe*

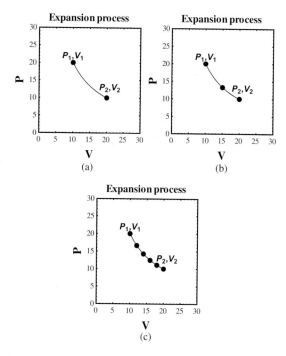

Figure N.1 Pressure Volume (PV) diagram. (a) Only the initial and the final points are shown, (b) one intermediate state added, (c) many intermediate states added.

system remains almost unchanged at P_1. It takes time to reach a new equilibrium state where the pressure will be the same at every point within the gas.

Looking at Figure N.1 we can say that the system has moved from the point P_1V_1 to the new point P_2V_2, but we do not know the *path* along which the system has changed in this diagram. In fact, there exists no path that describes the process in this diagram. The *volume* of the system is defined at each moment of the process, but the pressure of the system is undefined since along the process this system is not at equilibrium. Furthermore, we can reverse the process by placing a heavier weight on the piston. Since we do not know the path along which the system has moved in the PV diagram, it is not meaningful to talk about reversing the *path* of the process.

Next, suppose we begin at the same initial state as before, and that we replace the weight of mass M by a weight of mass $3M/4$ and wait until such time as the system reaches a new equilibrium state, say at $P_2 V_2$ (not the same as $P_2 V_2$ of the previous experiment), then replace the weight again with a weight of mass $M/2$ to reach a new equilibrium state, say, $P_3 V_3$, Fig. N.1b.

In this process we can describe the process of expansion from $P_1 V_1$ to $P_2 V_2$, and to $P_3 V_3$, but we cannot describe the states of the system at any intermediate point in the process. Therefore, we still cannot describe the *path* along which the system has moved from the initial to the final state. All we know is one intermediate point along this path.

Next, suppose that we start from an initial state $P_i V_i$, and reduce the weight on the piston from M to $9M/10$, wait for the equilibrium, reduce the weight from $9M/10$ to $8M/10$, and so on, until we have a weight of mass $M/10$. In such a process we can describe the path leading from $P_i V_i$ to the final state $P_f V_f$ by eight intermediary points in the PV diagram. We shall get a sequence of points, like the ones shown in Fig. N.1c looks almost like a "path". We can imagine even a slower process where at each step we reduce the pressure on the piston from P to $P - dp$ the work done by the system in this step is:

$$\delta W = -(P - dP)dV \approx -PdV$$

If we perform the expansion process in many small steps such that at each step we reduce the external pressure from P to P-dP, we shall obtain a dense sequence of equilibrium states. In the limit where the infinitesimal steps are so small that the equilibrium pressure of the system changes almost continuously we can describe the *path* of the process in the *PV* diagram as a continuous path leading from the initial to the final state. Along this path we know that the pressure is related to the volume (at constant temperature *T*) by:

$$P = \frac{nRT}{V}$$

Therefore, the work done by the system can be calculated by:

$$W\,(isothermal\ expansion) = -\int_{V_i}^{V_f} \frac{nRT}{V}\,dV = -nRT \ln \frac{V_f}{V_i}$$

We shall refer to the limiting process described above as a *quasi-static* process. A quasi-static process is a sequence of very small steps such that at every moment the equilibrium of the system changes almost continuously.

Clearly, such a process is only an idealization of a real process. In any real process, any change in the external pressure will cause a change in the internal pressure in the system. This change takes time. In the limit of very small steps we can say that the equilibrium pressure of the system changes almost continuously, and therefore a continuous path may be drawn in the *PV* diagram leading from the initial to the final state.

Having a path leading from the initial to the final state, it becomes meaningful to speak of *reversing* the process along the *same* path. Simply add very small weights on the piston and the process is reversed along the *same* path.

Because of the possibility of *reversing* the process along a well-defined path, a quasi-static process is also referred to as a *reversible* process. The term *reversible* is used in different senses; [see Ben-Naim (2011a)].

It should be noted that in this isothermal process of expansion of an ideal gas, the internal energy of the system does not change. This follows from the fact that at a constant temperature the average kinetic energy of the particles is unchanged. In this system the internal energy of the system is simply the total kinetic energy of the particles. Thus, from $\Delta E = 0$ and from the First Law we get:

$$\Delta E = Q + W = 0$$

Therefore, $$Q = -W = nRT \ln \frac{V_f}{V_i}$$

Thus, the amount of work done by the system (in lifting the weights) is equal to amount of heat that flowed from the bath into the system (note that heat entering the system is positive whereas work done by the system is negative).

Note 7: The exact quotation from Clausius is:

"Die energie der welt ist constant."
"Die entropie der welt strebt einem maximum zu."

The quotation given in the text is that which more commonly appears in the literature.

Note 8: $W(n)$ is calculated from $W(n) = \binom{N}{n} \frac{N!}{n!(N-n)!}$ and the probability $P_N(n)$ is given by: $P_N(n) = W(n)/2^N$. For more details, see Ben-Naim (2008, 2012, 2015).

Note 9: Suppose that the particles were numbered: 1, 2,..., N. We choose a number, say, 17, and ask what is the probability that this particular particle is in compartment L or R. The answer is x_L or x_R, respectively.

Note 10: Note that we deal here with ideal gases. If there are intermolecular interactions, then the final equilibrium state will not necessarily have a uniform density; one or more phases having different densities could co-exist. For instance, water at about 100°C and at one atmospheric pressure might have two phases at equilibrium, liquid and gas, having two different densities. Likewise, at the triple point of water we find three phases (i.e. solid, liquid and gas) at equilibrium, i.e. three different densities. The phase rule precludes the possibility of having more than three phases of a one-component system at equilibrium.

Note 11: This limit is imposed on us by the Heisenberg uncertainty principle. Here, we introduce the limit from the practical point of view.

Note 12: Previously, the Helmholtz energy was referred to as *free energy*, or Helmholtz free energy. Unfortunately, the word "free" in "free energy" was misinterpreted. It was misinterpreted even by those who warned their readers that "free" does not mean *energy for free*.

Note 13: The Legendre transformation.
The simplest Legendre transformation is from the energy function $E(S, V, N)$ to the Helmholtz energy function: $A(T, V, N)$. Note that in this transformation we replace one of the independent variables (S) by the derivative

$$T = \left(\frac{\partial E}{\partial S}\right)_{V,N} \tag{1}$$

One defines the new function

$$A(T, V, N) = E(T, V, N) - TS(T, V, N) \tag{2}$$

Note that in the last equation the chosen independent variables are (T, V, N). It can be shown that the extremum principle of the function $E(S, V, N)$ is "transformed" into an extremum principle for the function $A(T, V, N)$. Another important property of the Legendre transformation is its "reversibility," in the sense that having the function $A(T, V, N)$, one can transform back to E by defining the function

$$E(S, V, N) = A(S, V, N) + ST(S, V, N) \tag{3}$$

where

$$-S = \left(\frac{\partial A}{\partial T}\right)_{V,N} \tag{4}$$

Note that in equation (3) we chose the independent variables S, V, N. Note also the pattern in equations (2) and (3). We define a new function by taking the old function and subtracting the product of the two independent variables; the old and the new. The two equations (2) and (3) are the same except for the difference in the choice of independent variables.

The next Legendre transformation is from the function $A(T, V, N)$ to the Gibbs-energy function $G(T, P, N)$. Here, we replace the independent variable V, by P which is the derivative of A with respect to V.

$$-P = \left(\frac{\partial A}{\partial V}\right)_{T,N} \quad (5)$$

Thus, we define the function $G(T, P, N)$ as

$$G(T, P, N) = A(T, P, N) + PV(T, P, N) \quad (6)$$

Note that in equation (6) we chose the independent variables T, P, N to express all the other dependent variables.

As we have noted before the transformation from A to G can be reversed. We start with the function $G(T, P, N)$, take its derivative with respect to P

$$\left(\frac{\partial G}{\partial P}\right)_{T,N} = V \quad (7)$$

then define the new function

$$A(T, V, N) = G(T, V, N) - P(T, V, N)V \quad (8)$$

where now we chose the independent variables T, V, N.

What happens if we start from $G(T, P, N)$ and want to replace N by the chemical potential μ, which is defined by

$$\mu = \left(\frac{\partial G}{\partial N}\right)_{T,P} \quad (9)$$

Formally, we define a new function

$$f(T, P, \mu) = G(T, P, \mu) - \mu N(T, P, \mu) \qquad (10)$$

Clearly, this new function is identically zero. The reason is that for a one-component system we cannot choose the three intensive variables (T, P, μ) as independent variables. These variables are dependent. One way of expressing this dependence is by Gibbs–Duhem relationship

$$SdT - Vdp + Nd\mu = 0 \qquad (11)$$

which means that if you change two out of the three variables T, P, N, then the change in the third will be determined. This is equivalent to saying that there is an implicit function $f(T, P, \mu)$ which is identically zero

$$f(T, P, \mu) = 0 \qquad (12)$$

It should be noted that for a multi-component system we can perform a Legendre transformation, by replacing one or more (but not all) of the N_i by the corresponding chemical potential. For instance, we can start with the Gibbs energy function $G(T, P, N_1, ..., N_C)$ and replace the independent variable N_1 by the corresponding derivative

$$\mu_1 = \left(\frac{\partial G}{\partial N_1}\right)_{T, P, N_2, ..., N_C} \qquad (13)$$

Thus, we define the new function

$$F(T, P, \mu_1 N_2, ..., N_C) = G(T, P, \mu_1 N_2, ..., N_C) \\ - \mu_1 N_1(T, P, \mu_1 N_2, ..., N_C) \qquad (14)$$

Such functions are important for osmotic systems when one or more of the components can flow through a membrane which is permeable to one or more of the components.

In statistical mechanics the transformation of variables is achieved through the Laplace transform. For instance, starting with the entropy function $S(E, V, N)$ we can transform variables by either the sum or the integral

$$\sum_E \exp\left[-\frac{(E-TS)}{k_B T}\right], \quad \int \exp\left[-\frac{(E-TS)}{k_B T}\right] dE$$

In either case, by summing (or integrating) over the variable E, we get a quantity which is no longer a function of E, but a function of T.

Note14: The Gibbs energy was also referred to as free energy. As noted in Note 12, the word "free" might be misleading. The Gibbs energy is misunderstood even by those who warn the reader that free energy does not mean "energy is free." See for example, Carroll (2016).

Note 15: The arguments in this case are similar to the arguments in Section 4.6. Suppose we remove some constraint in the (T, V, N) system. Denote by $\Delta S(sys)$, $\Delta E(sys)$ and $\Delta V(sys)$, the change in entropy, energy and volume of the *system*. These can either be positive or negative.

Since the *total* is isolated, $\Delta S(total)$ must be positive and $\Delta E(total)$ must be zero.

From the first law we have, assuming only PV work;

$$\Delta E(sys) = \Delta W(sys) + \Delta Q(sys)$$
$$= -P\Delta V(sys) + \Delta Q(sys)$$

From the entropy formulation applied to the *total*:

$$\Delta S(tot) = \Delta S(sys) + \Delta S(bath) = \Delta S(sys) - \frac{\Delta Q(sys)}{T}$$
$$= \Delta S(sys) - \frac{1}{T}(\Delta E(sys) + P\Delta V(sys)) \geq 0$$

Re-arranging the last equation, we get

$$-T\Delta S(total) = \Delta E(sys) + P\Delta V(sys) - T\Delta S(sys) = \Delta G(sys) \leq 0$$

Remember that the *total* is isolated. Therefore, the change of entropy $\Delta S(total)$ must be positive. Therefore, $\Delta G(sys)$ must be negative.

Note 16: Strictly, we should deal with many solute particles at fixed positions $R_{i,s}$. However, for simplicity we deal with one solute particle only. For justification of this, see Ben-Naim (2006).

Note 17: The decrease in Gibbs energy is referred to the liberation Gibbs energy, it is $k_B T \ln \Lambda^3/V$, where Λ^3 is the momentum partition function arising from the Maxwell-Boltzmann distribution. For details, see Ben-Naim (1992, 2006).

Note 18: The Boltzmann distribution is $P(E) \approx \exp[-(E - E_0)/k_B T]$. This means that as the temperature approaches zero, the probability of finding the system at an energy $(E - E_0)$ above the lowest energy level E_0 (the ground state) becomes zero, except for the energy levels $E = E_0$.

Note 19: The average distance between pairs of particles will increase, but the interaction energy among all particles will go from repulsive to attractive, in going say, from x to f in Figure 5.3. The total potential energy will decrease. Therefore, if $\Delta(PE) < 0$ and since $\Delta E = 0$, $\Delta(KE) > 0$, which means that the temperature increases.

Note 20: Obviously, since in this case the average height of the particles will *decrease*, the total potential energy will decrease and the kinetic energy will increase, i.e. the temperature will increase.

References and Suggested Reading

Atkins, P. (2007), *Four Laws That Drive the Universe*, Oxford University Press.

Ben-Naim, A. (1987), Is Mixing a Thermodynamic Process? *Am. J. Phys.* 55; 725.

Ben-Naim, A. (1992), *Statistical Thermodynamics for Chemists and Biochemists*, Plenum Press, New York.

Ben-Naim, A. (2006a), *A Molecular Theory of Solutions*. Oxford University Press, Oxford.

Ben-Naim, A. (2006b), *American Journal of Physics.* 74; 1126.

Ben-Naim, A. (2007), *Entropy Demystified. The Second Law of Thermodynamics Reduced to Plain Common Sense.* World Scientific, Singapore.

Ben-Naim, A. (2008), *A Farewell to Entropy: Statistical Thermodynamics Based on Information*. World Scientific, Singapore.

Ben-Naim, A. (2009), An Informational-Theoretical Formulation of the Second Law of Thermodynamics. *J. Chem. Education*, 86; 99.

Ben-Naim, A. (2010), *Discover Entropy and the Second Law of Thermodynamics. A Playful Way of Discovering a Law of Nature*. World Scientific, Singapore.

Ben-Naim (2011a), *Molecular Theory of Water and Aqueous Solutions. Part II: The Role of Water in Protein Folding, Self-assembly and Molecular Recognition*. World Scientific, Singapore.

Ben-Naim, A. (2011b), Entropy: Order or Information. *J. Chem. Education*, 88; 594.

Ben-Naim, A. (2012), *Entropy and the Second Law. Interpretation and Misss-Interpretationsss*. World Scientific, Singapore.

Ben-Naim, A. (2013), *The Protein Folding Problem and Its Solutions*. World Scientific, Singapore.

Ben-Naim, A. (2014), *Statistical Thermodynamics, with Applications to Life Sciences,* World Scientific, Singapore.

Ben-Naim, A. (2015a), *Information, Entropy, Life and the Universe. What We Know and What We Do Not Know*. World Scientific, Singapore.

Ben-Naim, A. (2015b), *Discover Probability. How to Use It, How to Avoid Misusing It, and How It Affects Every Aspect of Your Life*. World Scientific, Singapore.

Ben-Naim, A. (2016a), *The Briefest History of Time*. World Scientific, Singapore.

Ben-Naim, A. (2016b), *Myths and Verities in Protein Folding Theories*. World Scientific, Singapore.

Ben-Naim, A. (2016c), *Entropy, the Truth the whole Truth and nothing but the Truth*. World Scientific, Singapore.

Ben-Naim A. and Casadei D. (2017) *Modern Thermodynamics*. World Scientfic, Singapore.

Bent, H.A. (1965), *The Second Law*. Oxford University Press, New York.

Boltzmann, L. (1877), *Vienna Academy*. 42; "Gesammelte Werke" p. 193.

Boltzmann, L. (1896), *Lectures on Gas Theory*. Translated by S.G. Brush, Dover, New York (1995).

Brillouin, L. (1962), *Science and Information Theory*. Academy Press, New York.

Brush, S. G. (1976), *The Kind Of Motion We Call Heat. A History Of The Kinetic Theory of Gases In The 19th Century, Book 2: Statistical Physics and Irreversible Processes*. North-Holland Publishing Company.

Brush, S. G. (1983), *Statistical Physics and the Atomic Theory of Matter, from Boyle and Newton to Landau and Onsager*. Princeton University Press, Princeton.

Callen, H.B. (1960), *Thermodynamics*. John Wiley and Sons, New York.

Callen, H.B. (1985), *Thermodynamics and an Introduction to Thermostatics*. 2nd edition. Wiley, New York.

Cooper, L. N. (1968), *An Introduction to the Meaning and Structure of Physics*. Harper and Low, New York.

Denbigh, K. (1981), *How Subjective id Entropy?* Chemistry in Britain, 17; 168.

Denbigh, K.G. and Denbigh, J.S. (1985), *Entropy in Relation to Incomplete Knowledge*. Cambridge University Press, Cambridge.

Denbigh, K.G. (1989), Note on Entropy, Disorder and Disorganization. *Brit. J. Phil. Sci.* 40; 323.

Dugdale, J.S. (1996), *Entropy and its Physical Meaning*. Taylor and Francis, London entropysite.oxy.com.

Fast, J.D. (1962), *Entropy. The Significance of the Concept of Entropy and its Applications in Science and Technology*. Philips Technical Library, Netherlands.

Fowler, R and Guggenheim, E.A. (1956) *Statistical Thermodynamics*. Cambridge University Press, Cambridge.

Gibbs, J.W. (1906), *Collected Scientific Papers of J. Willard Gibbs*. Longmans, Green New York.

Hill, T.L. (1960), Introduction.

Tribus M. and McIrvine, E.C. (1971), *Entropy and Information*. Scientific American, 225; 179.

Wilks J. (1961) *The Third Law of Thermodynamics*. Oxford University press, Oxford.

Index

Adiabatic
   cooling, 143
   expansion, 143
   expansion in gravitational field, 145–147
   demagnetization, 147–149
   system, 3–4
Avogadro number, 2

Conservation of energy, 54–61
Conservation of mass-energy, 61–62

Entropy
   absolute value, 138–142
   and disorder, 71, 99
   and information theory, 73
   and "quality of energy", 71
   Boltzmann's definition, 97
   Clausius definition, 96
   definition based on SMI, 96–99
   residual, 138, 142
Equilibrium state, 11–16
   thermal, 42
Extensive parameter, 5–7

First Law, 53–56
   definition, 59–61

Gibbs energy, 115–118
   and the Second Law, 115
   as "disguised" entropy, 117
   definition of, 115

Heat flow, 58–59
Helmholtz energy, 107–114
    and Second Law, 107–109, 113
    as "disguised" entropy, 112
    definition, 107
Hess' Law, 63–64

Indistinguishability, 36
Intensive parameter, 7–10
Isobaric system, 4
Isolated system, 3
Isothermal system, 4
Isothermal isobaric system, 4

Legendre transformation, 169–171

Noether's theorem, 65–66

Probability distribution, 75–85

Quasi-static process, 18

Residual entropy, 138–142

Shannon's measure of information, 25–39
    and entropy, 37–38, 94, 96–99
    and twenty-question game, 28–38
    definition, 32, 96
State function, 16–18

The Second Law, 69–133
    entropy formulation, 93–106, 167
    Gibbs energy formulation, 115–118
    Helmholtz energy formulation, 106–114
    Probability formulation, 85–93

Work, 57–58, 161–163

Zeroth Law, 41–51, 159
    material, 48–50
    mechanical, 47–48
    thermal, 41–47

Other Recent Books by the Author

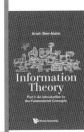

Information Theory
Part I: An Introduction to the Fundamental Concept

By: Arieh Ben-Naim

ISBN: 978-981-3208-82-7
ISBN: 978-981-3208-83-4 (pbk)

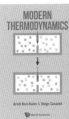

Modern Thermodynamics

By: Arieh Ben-Naim &
Diego Casadei

ISBN: 978-981-3200-75-3
ISBN: 978-981-3200-76-0 (pbk)

Entropy
The Truth, the Whole Truth, and Nothing But the Truth

By: Arieh Ben-Naim

ISBN: 978-981-3147-66-9
ISBN: 978-981-3147-67-6 (pbk)

The Briefest History of Time
The History of Histories of Time and the Misconstrued Association between Entropy and Time

By: Arieh Ben-Naim

ISBN: 978-981-4749-84-8
ISBN: 978-981-4749-85-5 (pbk)

Entropy Demystified
The Second Law Reduced to Plain Common Sense
2nd Edition

By: Arieh Ben-Naim

ISBN: 978-981-3100-11-4
ISBN: 978-981-3100-12-1 (pbk)

Myths and Verities in Protein Folding Theories

By: Arieh Ben-Naim

ISBN: 978-981-4725-98-9
ISBN: 978-981-4725-99-6 (pbk)

Information, Entropy, Life and the Universe
What We Know and What We Do Not Know

By: Arieh Ben-Naim

ISBN: 978-981-4651-66-0
ISBN: 978-981-4651-67-7 (pbk)

Discover Probability
How to Use It, How to Avoid Misusing It, and How It Affects Every Aspect of Your Life

By: Arieh Ben-Naim

ISBN: 978-981-4616-31-7
ISBN: 978-981-4616-32-4 (pbk)

Statistical Thermodynamics

With Applications to
the Life Sciences

By: Arieh Ben-Naim

ISBN: 978-981-4579-15-5
ISBN: 978-981-4578-20-2 (pbk)

Alice's Adventures in Molecular Biology

By: Arieh Ben-Naim &
Roberta Ben-Naim

ISBN: 978-981-4417-24-2
ISBN: 978-981-4417-25-9 (pbk)

The Protein Folding Problem and Its Solutions

By: Arieh Ben-Naim

ISBN: 978-981-4436-35-9
ISBN: 978-981-4436-36-6 (pbk)

Entropy and the Second Law
Interpretation and
Misss-Interpretationsss

By: Arieh Ben-Naim

ISBN: 978-981-4407-55-7
ISBN: 978-981-4374-89-7 (pbk)

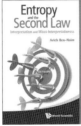

World Scientific
Connecting Great Minds

Alice's Adventures in Water-land

By: Arieh Ben-Naim & Roberta Ben-Naim

ISBN: 978-981-4338-96-7 (pbk)

Molecular Theory of Water and Aqueous Solutions

Part II: The Role of Water in Protein Folding, Self-Assembly and Molecular Recognition

By: Arieh Ben-Naim

ISBN: 978-981-4383-11-0 (Set)
ISBN: 978-981-4383-12-7 (pbk) (Set)
ISBN: 978-981-4350-53-2
ISBN: 978-981-4350-54-9 (pbk)

Discover Entropy and the Second Law of Thermodynamics

A Playful Way of Discovering a Law of Nature

By: Arieh Ben-Naim

ISBN: 978-981-4299-75-6
ISBN: 978-981-4299-76-3 (pbk)

Molecular Theory of Water and Aqueous Solutions

Part I: Understanding Water

By: Arieh Ben-Naim

ISBN: 978-981-283-760-8
ISBN: 978-981-4327-71-8 (pbk)

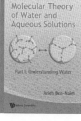